# Ancient West Asian Civilization

Akira Tsuneki • Shigeo Yamada
Ken-ichiro Hisada
Editors

# Ancient West Asian Civilization

## Geoenvironment and Society in the Pre-Islamic Middle East

Springer

*Editors*
Akira Tsuneki
Faculty of Humanities and Social Sciences
University of Tsukuba
Tsukuba, Ibaraki, Japan

Shigeo Yamada
Faculty of Humanities and Social Sciences
University of Tsukuba
Tsukuba, Ibaraki, Japan

Ken-ichiro Hisada
Faculty of Life and Environmental Sciences
University of Tsukuba
Tsukuba, Ibaraki, Japan

ISBN 978-981-10-0553-4 ISBN 978-981-10-0554-1 (eBook)
DOI 10.1007/978-981-10-0554-1

Library of Congress Control Number: 2016949386



Printed on acid-free paper

This Springer imprint is published by Springer Nature
The registered company is Springer Science+Business Media Singapore Pte Ltd.

# Preface

The publication of this book is closely related to the symposium "Facilitating the Study of West Asian Civilization: What Does Ancient West Asia Tell Us?" held in Tokyo June 28–29, 2014. The symposium aimed to achieve a wide perspective on West Asian environments and civilizations while gathering the collective knowledge of geologists, archaeologists, philologists, and historians. All of the chapters in the present volume were written by symposium participants. The authors reviewed their work after discussion and exchange of information. Consequently, their contributions reflect the central theme of this book: understanding the essence of West Asian civilization, its geological and geographical foundation, and its significant influence on the modern world.

The editors wish to thank all the authors, as well as other colleagues who gave presentations at the symposium and supported the contents and structure of this book. Publication was carried out as part of an ongoing research project of the Research Center for West Asian Civilization, University of Tsukuba (Ibaraki, Japan), which is financially supported by grants from the Japanese Ministry of Education, Culture, Sports, Sciences and Technology (MEXT). We are grateful to the staff of the Research Center, which unfailingly assisted us in countless ways.

In today's immense political turmoil, the Middle East often tends to be talked about for its negative aspects via negative images. However, this region has contributed to the foundation of elements necessary in all modern human societies. Therefore, we hope that the importance of Ancient West Asia is realized again, and if this book contributes to this purpose, we are content.

Tsukuba, Ibaraki, Japan

Akira Tsuneki
Shigeo Yamada
Ken-ichiro Hisada

# Contents

# Chapter 1
# Introduction: The Aim of *Study of Ancient West Asian Civilization*

**Akira Tsuneki**

## 1.1 The Terms *West Asia* and *Ancient West Asian Civilization*

This book was planned to facilitate a new discipline, the *Study of Ancient West Asian Civilization*. First of all, we must explain the terms "West Asia" and "ancient West Asian civilization". The term "West Asia" is used for the study area in this book and indicates the western part of Asia, which has generally been referred to as the Middle East. The term Middle East (Near East was gradually replaced by the term Middle East after World War II) was first used by European and American politicians and military personnel in the late nineteenth century (Beaumont et al. 1976; Koppes 1976). The designation of the part of Asia west of India, as the "Middle East" was convenient, and this word has been indispensable in discussing political and historical issues, such as "the Middle Eastern question" (Chirol 1903). However, this word is value-laden and represents a substantial eurocentric viewpoint. For example, from the viewpoint of Japan, the Middle East must be termed the Far East as it is the furthermost region in Asia (East). We would like to exclude this eurocentric synonym (Middle East) and mitigate potential bias (i.e. historical and Western) by using a purely geographical term. Today, the term West/Western Asia is normally used by non-Western-centralized countries, magazines, and international institutes, such as *National Geographic* (2006) and the United Nations (2013).

West Asia in this book is identical to the narrower Middle East, which encompasses modern Iran and Turkey from east to west and Turkey and the Arabian Peninsula from north to south. West Asia excludes Egypt, which is traditionally contained in the term Middle East. Ancient Egypt will not be included in the scope of this discussion because we consider the early history of Egypt developed under a different philosophy and entity from those of West Asia though both civilizations developed concurrently.

A. Tsuneki (✉)
Faculty of Humanities and Social Sciences, University of Tsukuba,
1-1-1, Tennodai, Tsukuba 305-8571, Ibaraki, Japan
e-mail: tsuneki.akira.gf@u.tsukuba.ac.jp



A. Tsuneki et al. (eds.), *Ancient West Asian Civilization*,
DOI 10.1007/978-981-10-0554-1_1

"Ancient West Asian civilization" indicates the body of cultures which appeared in West Asia mainly between the 10th and 1st millennia BC. As we will discuss in detail later, ancient West Asian civilization provided the foundations for all modern civilizations, from fundamental foodstuff and technological innovation to social systems including spiritual life. This is the crucial point. For understanding the formation and future of modern civilizations, the *Study of Ancient West Asian Civilization* is indispensable. This is the reason why we would like to facilitate research in this new discipline.

## 1.2 The Necessity of New Perspectives from Ancient West Asian Civilization

A series of great transformations in human history occurred in ancient West Asia however these facts tend to be forgotten by people living in the modern world. The religions, politics, economies, and cultures of West Asia are seen as being in conflict, producing an unstable social climate. The region has been treated as a troublemaker in modern world politics. In particular, the Islamic societies of West Asia have been turned into scapegoat. Islamic societies have also become counterpoints to Western societies and have become symbolic of anti-Western viewpoints. *The Clash of Civilizations and the Remaking of World Order* by Samuel Huntington (1996), has deeply affected modern politics, and suggests that the main conflicts after the Cold War changed from struggles between nations to those between civilizations. Huntington predicted that the world order would ultimately reach an impasse. Instead of national borders and flags, people start to assemble on behalf of their own culture, civilization and identity. One of the most severe axes of conflict is between Western (European and North American) and Islamic civilizations. People began to lay great emphasis on cultural differences and this has accelerated after September 11, 2001.

However, when we examine modern issues from the viewpoint of ancient West Asian civilization it is necessary to determine if West European civilization is in opposition to Islamic civilization and whether each civilization has grown to be incompatible with the other. Most of the fundamental characteristics of modern Islamic societies (e.g., the interpersonal relationships based on kinship, and the strong belief in monotheism) had appeared and developed in West Asia long before the appearance of Islam. Modern Islamic societies, whether Arab or not, cannot be understood without a deep understanding of ancient West Asian civilization. Once familiar with ancient West Asian civilization it is possible to identify the primary elements, such as wheat cultivation, urbanism and Christianity, which are also fundamental in Western (European and North American) societies that have originated in ancient West Asian civilization. Modern Islamic and Western civilizations are sibling civilizations that have their roots in ancient West Asian civilization. Therefore, the study of ancient West Asian civilization contributes to the understanding of not only modern Islamic and other West Asian societies but also of the foundation of Western societies. The *Study of Ancient West Asian Civilization* must contribute to and promote mutual understanding between these two civilizations. As most modern civilizations owe many of their elements to Western civilizations, we believe that the *Study of Ancient West Asian Civilization* is indispensable in understanding the foundations of the modern world.

We hope to create a more mutually understandable world through the *Study of Ancient West Asian Civilization*. Of course this path is new and appears to be rudimentary however, we hope that the publishing of this book encourages and advances this new field.

## 1.3 Natural Environment and Its Significance for Civilization

West Asia is located at the intersection of the African, Asian and European continents (Fig. 1.1). It has provided the corridors for transmigration, cultural interactions, economic trade, military advances, evangelization, and other human activities. Its geographical location elevated West Asia to the position of a focal point for the dynamic transformations in human history. If we compare this condition with the impasse conditions experienced by the British Isles or the Japanese Archipelago, the difference becomes definitive. In the latter, human groups and cultures immigrated into areas then lacked exit routes and populations mixed repeatedly and densely. Human and cultural imports were excessive compared to exports before modernization. This included varying periods of isolation. On the contrary, people in West

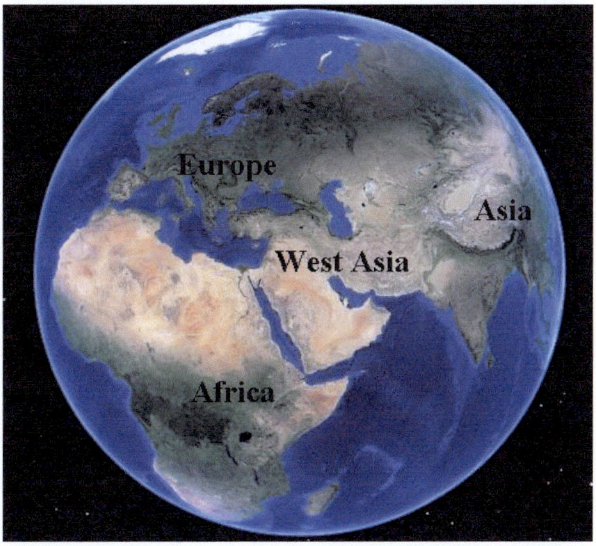


**Fig. 1.1** West Asia connecting the three continents (After Google Earth)

Asia constantly engaged with other human groups. Their land was open to all and literally a crossroads for humans and cultures. Under such conditions, their history must have been dynamic whether the people of West Asia liked it or not.

West Asia also has extensive natural and mineral resources. Today the world economy is and has been controlled for a long period of time by the petroleum policy of West Asian countries and international capitals. This phenomenon is not limited to modern times. In the prehistory and early history of West Asia, mineral resources, such as obsidian, copper, tin, lead, gold, silver, iron, and precious stones, played important roles in the development of towns, cities and civilization. The polities of early settlements and states tried to control natural resources, including minerals, and used them to dominate other human groups. Ken-ichiro Hisada and Seyed Aghanabati explain about the abundant mineral resources of ancient West Asia in Chaps. 2 and 3. Modern West Asia was favored with petroleum and Ancient West Asia with mineral resources useful for human beings.

Paleoclimatic changes and the evolution of human cultures in ancient West Asia is discussed by Ryo Anma and Teruyuki Maruoka in Chap. 4. They consider some specific examples of geologically driven environmental changes in natural-cultural landscapes and evaluate their impacts on civilization, as well as the implications for societies and nations at present and in the future. The effects caused by sudden environmental changes (natural disasters) and seismicity on human societies, is discussed by Tsuneo Ohsumi and and Yuji Yagi in Chap. 5.

## 1.4 Societal Elements and Technologies That Originated in Ancient West Asia

There are many familiar social and technological elements which originated in ancient West Asia. Some major foodstuffs first cultivated and developed in West Asia include: wheat, barley, and grapes, which are the ingredients for bread, beer and wine, beans (e.g., peas and lentils), meats (e.g., beef, pork and mutton), dairy products (e.g., cheese and yogurt), fruits and vegetables (e.g., onion, carrot, spinach and apple), olives and olive oil, and coffee, etc. All of these modern foodstuffs originated and were developed in West Asia. Of course, foodstuffs that originated in various regions of the world were exploited and became important modern foodstuffs for human beings. However, considering the earliness of exploitation and the importance of major human foodstuffs, those originating in West Asia are conspicuous (i.e. Larson et al. 2014). Our modern world cannot be maintained without these West Asian foodstuffs. In Chaps. 6 and 7, Ken-ichi Tanno, Osamu Maeda, Hans-Peter and Margarethe Uerpmann discuss the origin and development of major food sources exploited in West Asia using the latest data.

The fundamental technologies and social systems present in the modern world also originated in ancient West Asia. Today, iron is still the most important basic material, and the technologies which initially produced iron can be traced to ancient

West Asian civilization. Copper, then bronze, and later iron technologies were developed in succession in ancient West Asia. Many kinds of weapons, tools, and other equipment were produced from these metals, and they played important roles within societies. Technological advancements in ancient West Asia also included glass working and various types of metallurgy. West Asia was the most advanced region in the world with regard to technological developments between the 10th and 1st millennia BC. Moreover, these technologies, advanced for their time, provide the building blocks of today's modern technologies. In Chap. 8 ÜnsalYalçin discusses how and when people started metal working in prehistoric West Asia.

## 1.5 Toward Urbanization

Social systems that were developed in ancient West Asia also affected life in the modern world. Currently, over half the world's population lives in urban communities (United Nations 2015). Urbanism first developed in West Asia, as did the social systems (i.e., states) which governed urban communities. To understand the origin and development of urban communities and state societies, it is necessary to examine ancient West Asian history. The *Study of Ancient West Asian Civilization* reveals the essence of cities and states, and allows us to consider why people gather in cities. This is particularly relevant to issues tied to modern urban problems. Jason Ur discusses the origin of cities using results from remote-sensing and proposes the latest hypothesis for the urbanization process in Chap. 9.

So many social innovations appeared within early cities and states. The most representative is a literary system, which is integral to systems of communication within and between persons and organizations. The first system of writing was invented in ancient West Asia, and then spread throughout the Old World. While some scholars assert that writing was invented in multiple locations (e.g. Daniels and Bright 1996:2), others believe that the complicated symbolic systems used for writing could not have been invented independently in so many locations and times. They suggest that the idea of writing, but not the particular symbols of a script, originated from the oldest writing system, i.e. Mesopotamian cuneiform (e.g. Robinson 1995:12). Nevertheless, proto-writing pictograms first appeared around 3200 BC in Mesopotamia. From these pictograms the first complete system of writing, cuneiform, was invented by 3100 BC. Study of pictograms and cuneiform clay tablets shows that there can be no question of the first writings in West Asia being invented for the administration of goods. In Chaps. 10 and 11, Jonathan Taylor, Jun Ikeda and Shigeo Yamada introduce and challenge the process for the invention and development of cuneiform letters.

The evidence for the administration of goods extends back into the Neolithic sealing systems with the use of tokens and stamp seals in West Asia. Therefore, 4000 years of history leading up to the development of writing systems in West Asia can be traced to these sealing systems. The author excavated a Neolithic mega site named Tell el-Kerkh in northwest Syria from 1997 (Tsuneki et al. 2007; Tsuneki

**Fig. 1.2** Neolithic stamp seals found during the excavations at Tell el-Kerkh

2012). This site is one of the sites that produced numerous Neolithic stamp seals (Fig. 1.2). We discovered over one-hundred stamp seals and six clay sealings mainly from the middle 7th millennium BC cultural layers, which tells us about the early stationery used for goods-administration. Five of six clay sealings have a seal impression on the obverse surface and traces of strings and basketry on the reverse surface (Fig. 1.3). It is certain that they were used for sealing containers, such as baskets (Fig. 1.4 upper). Another clay sealing has a circular seal impression with a bucranium-like design on the obverse surface (Fig. 1.3 low-right). Its reverse surface is plain without any container impression, but traces of a string and knot can be observed on its broken side. Thus, it is assumed that this sealing was not attached directly to the containers, but rather hung from the container or other object like a label (Fig. 1.4 lower). Therefore, this object represents the transition from sealing to record-keeping devices. As we know, clay balls developed into clay bullae, then, into clay tablets in the Uruk period (Fig. 1.5). So, the Kerkh seals and sealings provide us with evidence for the early development of a goods-administration system and a long precedent for the development of writing systems in West Asia. This long sequence of evidence demonstrates that writing systems were certainly invented independently in West Asia. If the history of West Asian writing is compared to Egyptian hieroglyphs and Indus scripts, which lack pre- and proto-writing and appeared suddenly as complete writing systems, the unique local development of writing in West Asia is significant. Later on, alphabets evolved from the writings of

**Fig. 1.3** Neolithic clay sealings found during the excavations at Tell el-Kerkh

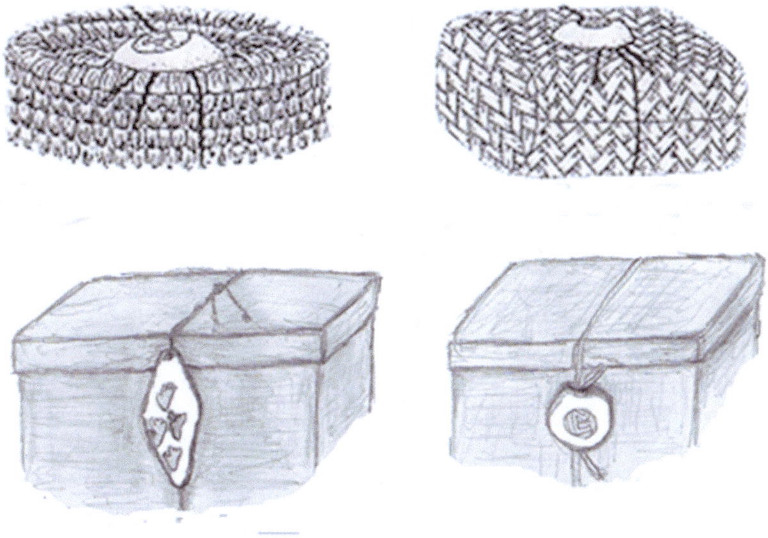

**Fig. 1.4** Presumed usage of Neolithic clay sealings

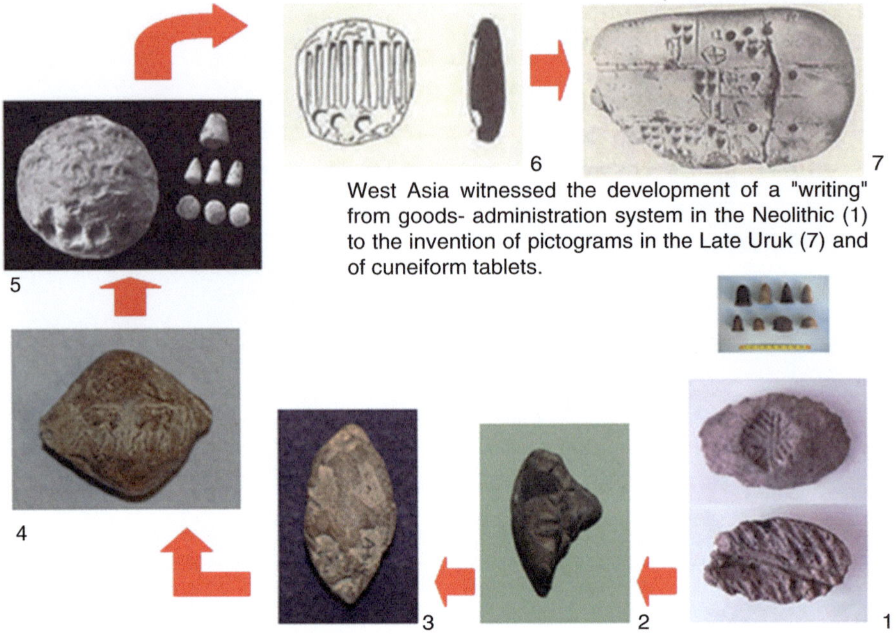


**Fig. 1.5** Development process of writing. *1–2*: Tell el-Kerkh clay sealings, *3*: Tell Arpachiyah clay bulla, *4*: Nineveh clay bulla, *5*: Late Uruk clay bulla (after Robinson 1995:61), *6*: Susa numerical tablet (after Vallat 1986), *7*: Late Uruk pictogram tablet (after Schmandt-Besserat 1992)

the late 2nd millennium BC in the eastern Mediterranean region, and then dispersed around the world.

The contribution of West Asian civilization to modern human history is reflected in material culture and social systems as well as in benevolent and spiritual matters. At present, followers of monotheistic religions, such as Christianity and Islam, comprise the majority of religious individuals in the world (Pew Research Center 2012). Everyone knows that Islam has a direct connection to Christianity and Judaism. Allah/Yahwe originated from the divine of ancient West Asia. There are various hypotheses for the origin of the name Yahwe. Some scholars assert that the title came from “El”, the head of the Cannanite pantheon (Chalmers 2012), others from a storm god (Smith 2000, 2001), and others from Egyptian inscription, the land name of “Yahu (Grabbe 2007)”. It is apparent that the gods and goddesses depicted in the West Asian pantheon were worshipped by ancient people and these gods and goddesses probably date back to the Early Dynastic period or earlier. Here again, is evidence of a very long history of belief derived from ancient West Asia and present in the modern world. In Chap. 12, Uri Gabbay discusses mythology, cult and scripture of ancient Mesopotamian polytheistic religion, and directs us to consider the appearance of monotheism, a belief which deeply affects the modern world.

## 1.6 Aim of This Book

This book is the first step towards creating a new discipline, the *Study of Ancient West Asian Civilization*. West Asia has been the focal point of great inventions and transformations in human history. It was the dispersal point for *Homo ergaster* and *Homo sapiens* when they departed from Africa to other continents. From the 10th to 1st millennia BC, this region is extremely important in the history of innovation, such as the origin of agriculture, development of metallurgy, formation of cities, invention of writing, organization of territorial states and the appearance of monotheism. West Asia was technologically and socially advanced compared to the rest of the world during this ten thousand year-period of human history. Various groups transmitted new innovation from West Asia to neighboring regions (e.g., Egypt, South Asia, Central Asia, and Europe) at different times. These innovations took root in each region and were improved, refined, and spread throughout the world, particularly by European people in early modern times. These West Asian innovations have had a significant impact on the rest of the world and provided the foundations for modern civilizations.

Therefore, this book tries to recognize West Asia as the place where the foundation for all modern civilizations was created. We discuss the relationship between the natural environment and civilization (Part 1) from which people created the earliest civilization in the world unrivalled elsewhere. We explain when, how, and why a series of great transformations and innovations appeared in ancient West Asia (Parts 2 and 3). We have undertaken fieldwork in many locations in West Asia, collected valuable data, and promote our research on ancient West Asian civilization in this publication. Most chapters in this book are the results of fieldwork in West Asia. We believe that these studies provide the opportunity to reconsider how we should live a better and more profound life. At the same time, reconsideration of the people and culture of ancient West Asia will facilitate true respect for their achievements in each period and area.

Unfortunately, we now face a cultural heritage crisis in West Asia, particularly in Iraq and Syria. So, we would like to continue our efforts to not only search for the truth but also to protect cultural heritage which is the witness to history. Yoko Taniguchi, Stavroula Golfomitsou, Thilo Rehren and Christian Eckmann discuss the roles and possibilities of archaeological science for the preservation of cultural heritage in Part 4, Chaps. 13 and 14.

In the synthesis Timothy Harrison summarizes the great transformations in human history. He argues that in order to study historical transformations related to modern world problems, we must examine the methods for studying civilizations and social developments, such as the "long-term lock-in theory", "short-term accidental theory", and "social development theory". When we study transformations over a long period, we must endeavor to look for more flexible methods such as viewpoints encompassing environmental change, subsistence, social struggles and cultural interaction (soft power). In addition to a deep-time perspective, he expects the increase in Asian and non-Eurocentric perspectives will further develop new civilization studies. He considers research on cultural interaction will be the key for understanding great transformations.

In conclusion, the three editors of this book extracted the background conditions for a series of great transformations and innovations between the 10th and 1st millennia BC. We recognize again the extreme richness of the natural resources of West Asia, in particular useful minerals, and domesticated animals and plants. The diversity in the natural environment and mutual interaction between human groups was remarkable in West Asia. These natural and cultural conditions promoted advancement and universality in West Asian civilization. We must continue to consider why ancient West Asian civilization created and developed so many important innovations which formed the foundation of modern civilizations. We endeavor to create a new discipline, the *Study of Ancient West Asian Civilization*, to focus on the history of West Asia prior to Islam. In promoting this new discipline, we build a new image of West Asia, which can stand in opposition to the prejudice and negative image that is currently commonplace.

## References

Beaumont, P., G.H. Blake, and J.M. Wagstaff. 1976. *The Middle East: A geographical study*. London: John Wiley & Sons.

Chalmers, A. 2012. *Exploring the religion of ancient Israel: Prophet, priest, sage and people*. London: SPCK.

Chirol, V. 1903. *The Middle Eastern question or some political problems of Indian defence*. London: John Murray

Daniels, P.T., and W. Bright (eds.). 1996. *The world's writing systems*. Oxford: Oxford University Press.

Grabbe, L. 2007. *Ancient Israel: What do we know and how do we know it?* New York: T&T Clark.

Huntington, S. 1996. *The clash of civilizations and the remaking of world order*. New York: Simon & Schuster Inc.

Koppes, C.R. 1976. Captain Mahan, General Gordon and the origins of the term "Middle East". *Middle Eastern Studies* 12-1: 95–98.

Larson, G., D.R. Piperno, R.G. Allaby, M.D. Purugganan, L. Andersson, M. Arroyo-Kalin, L. Barton, C.C. Vigueira, T. Denham, K. Dobney, A.N. Doust, P. Gepts, M.T.P. Gilbert, K.J. Gremillion, L. Lucas, L. Lukens, F.B. Marshall, K.M. Olsen, J.C. Pires, P.J. Richerson, R.R. de Casas, O.I. Sanjur, M.G. Thomas, M.J. West-Eberhard, and D.Q. Fuller. 2014. Current perspectives and the future of domestication studies. *PNAS* 111(17): 6139–6146.

National Geographic. 2006. *National Geographic style manual*. http://stylemanual.ngs.org/home/A/asian.

Pew Research Center. 2012. *The global religious landscape*. http://www.pewforum.org/2012/12/18/global-reli-gious-landscape-exec/.

Robinson, A. 1995. *The story of writing*. London: Thames and Hudson.

Schmandt-Besserat, D. 1992. *Before writing: From counting to Cuneiform*. Austin: University of Texas Press.

Smith, M.S. 2000. El. In *Eerdmans dictionary of the Bible*, ed. D.N. Freedman, A.C. Myers, and A.B. Beck. Cambridge, MA: Eerdmans.

Smith, M.S. 2001. *The origins of biblical monotheism: Israel's polytheistic background and the Ugaritic texts*. Oxford: Oxford University Press.

Tsuneki, A. 2012. Tell el-Kerkh as a Neolithic mega site. *Orient* 47: 29–65.

Tsuneki, A., J. Hydar, T. Odaka, and A. Hasegawa. 2007. *A decade of excavations at Tell el-Kerkh, 1997–2006*. Tsukuba: Department of Archaeology, University of Tsukuba.
United Nations. 2013. *United Nations statistics division*. http://unstats.un.org/unsd/methods/m49/m49regin.htm.
United Nations. 2015. *World urbanization prospects*. http://esa.un.org/unup/.
Vallat, F. 1986. The most ancient scripts of Iran: The current situation. *World Archaeology* 17–3: 335–347.

# Part I
# Environment and People of Ancient West Asia

# Chapter 2
# Geology Based Culture?

**Ken-ichiro Hisada**

The relationship between mineral resources and humans started when ancient people picked up suitably shaped stones for use as hunting aids. They discovered that siliceous nodules, chert, radiolarite, obsidian, and other hard rocks made the best knives and spear points. Native metals such as copper and gold were first used more than 20,000 years ago (Skinner et al. 2013). In ca. 5000 BC ancient people began to learn how to extract copper from certain minerals through a process known as smelting, which will be described in Chap. 8 in this volume. They found how to smelt minerals to produce lead, tin, zinc, silver, and other metals. The technique of mixing metals to create alloys was learned next, and bronze came into use by combining copper and tin. In this chapter, continental and regional topography and geological features of West Asia will be first described to discuss the relationship between the geology and the people.

## 2.1 Outline of West Asia

West Asia is located east of the Mediterranean Sea and west of Central Asia and South Asia. It is separated from Europe by the Bosporus strait and from Africa by the Red Sea (Fig. 2.1). According to the division of climatic zones by Köppen, most of West Asia belongs to dry desert and steppe climates. The coastal areas along the Mediterranean Sea, the Black Sea, and the Caspian Sea belong to extratropical

K.-i. Hisada (✉)
Faculty of Life and Environmental Sciences, University of Tsukuba,
1-1-1 Tennodai, Tsukuba 305-8572, Ibaraki, Japan
e-mail: hisadak@geol.tsukuba.ac.jp

 
A. Tsuneki et al. (eds.), *Ancient West Asian Civilization*,
DOI 10.1007/978-981-10-0554-1_2

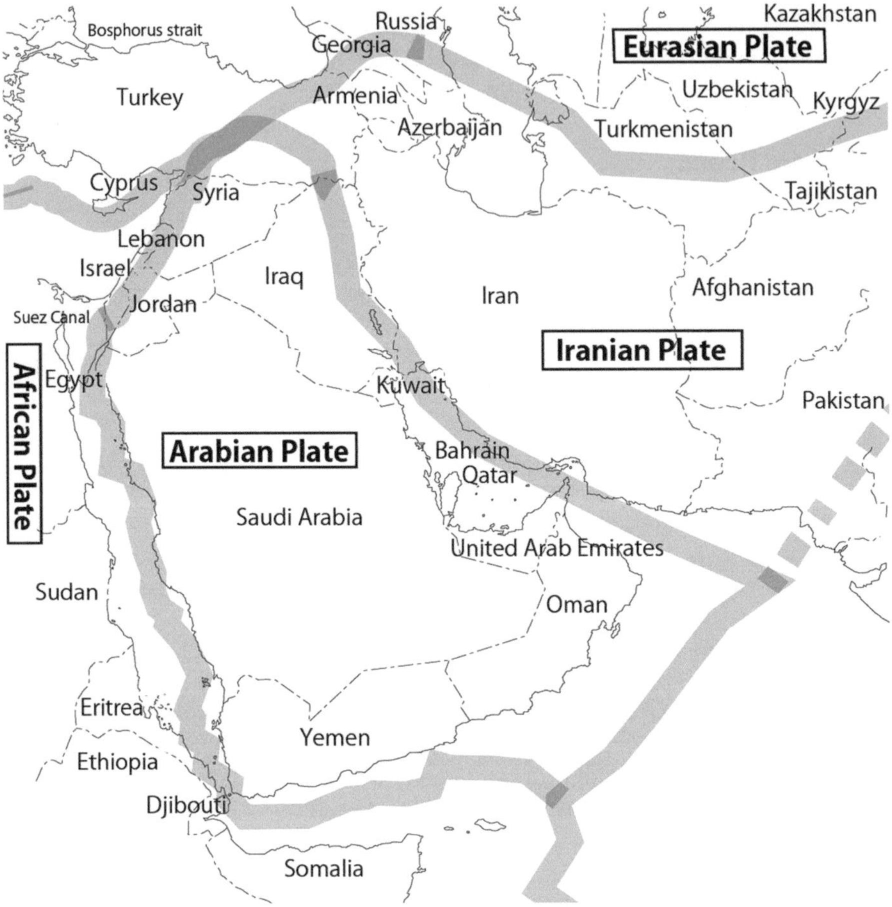


**Fig. 2.1** Plate boundaries of West Asia

Mediterranean, highlands Mediterranean and subarctic wet climates. It is obvious that these climates had a considerable influence on civilization and the cultural formation of these lands. Such climatic zone formation, however, can be traced back to plate tectonics, which the created topographic relief of West Asia such as high mountains along mobile belts and flatstable landmasses.

## 2.2 Plate Tectonics of West Asia

The outer layer of the earth is made of the firm, rigid lithosphere above the softer asthenosphere, where materials can easily flow. Much like a person slipping on a banana skin, the lithosphere easily glides over the asthenosphere. The lithosphere is divided into oceanic plates and continental plates moving in various directions to form three different types of boundaries. Convergent boundaries occur when plates collide. In such cases, the collision force can result in buckling of the two plates, or one plate may slide beneath the other. Divergent boundaries occur when the plates are pulled apart, and transform plate boundaries occur when two plates slide laterally against each other.

The thickness of the plates has been estimated to be more than several tens of kilometers to 150 km. Continent-scale large topography of West Asia was formed by movement of four plates: the Eurasian, Iranian, Arabian, and African plates (Fig. 2.1). As previously mentioned, plate boundaries create mobile belts, along which volcanoes and earthquakes often occur. Such activities in fact, trace and define the shape of the plate boundary. The characteristics of the plate boundaries in West Asia are described below.

(a) The Arabian and Iranian plates formed a collision zone. The collision caused orogenic movement and built up the Zagros Mountains.
(b) The border of the Arabian and Eurasia plates formed a collision zone. Near Cyprus in the Mediterranean Sea, however, the convergence formed a subduction zone.
(c) The border between the African and Arabian plates exhibits different characteristics throughout the region. Divergent and transform boundaries occur in the south and north regions, respectively.

## 2.3 Continental and Regional Topography of West Asia

### *2.3.1 General View of West Asian Topography*

The Arabian Peninsula lies in the southern region of West Asia. To the north, mountain ranges such as the Pontic Mountains, Caucasus Mountains, Alborz Mountains, and Zagros Mountains run northwest to southeast, and deserts and high plateaus extend among these mountain ranges (Fig. 2.2). This great mountain range is located between the European Alps and the Himalayas. Its highest peak is Mount Damavand (5671 m), which is a stratovolcano situated east of Tehran.

The Arabian Peninsula is surrounded by the Mediterranean Sea, Red Sea, Strait of Bab-el-Mandeb, Gulf of Aden, Strait of Hormuz, Gulf of Oman, and the Arabian Sea (Fig. 2.2). The northern part of the Arabian plate forms the basement to the Fertile Crescent (Fig. 2.3), and the Euphrates and the Tigris rivers flow through it. Both rivers join into Shatt al-Arab River and flow into the Gulf between the Arabian peninsular and Eurasian continent. Ancient Mesopotamian

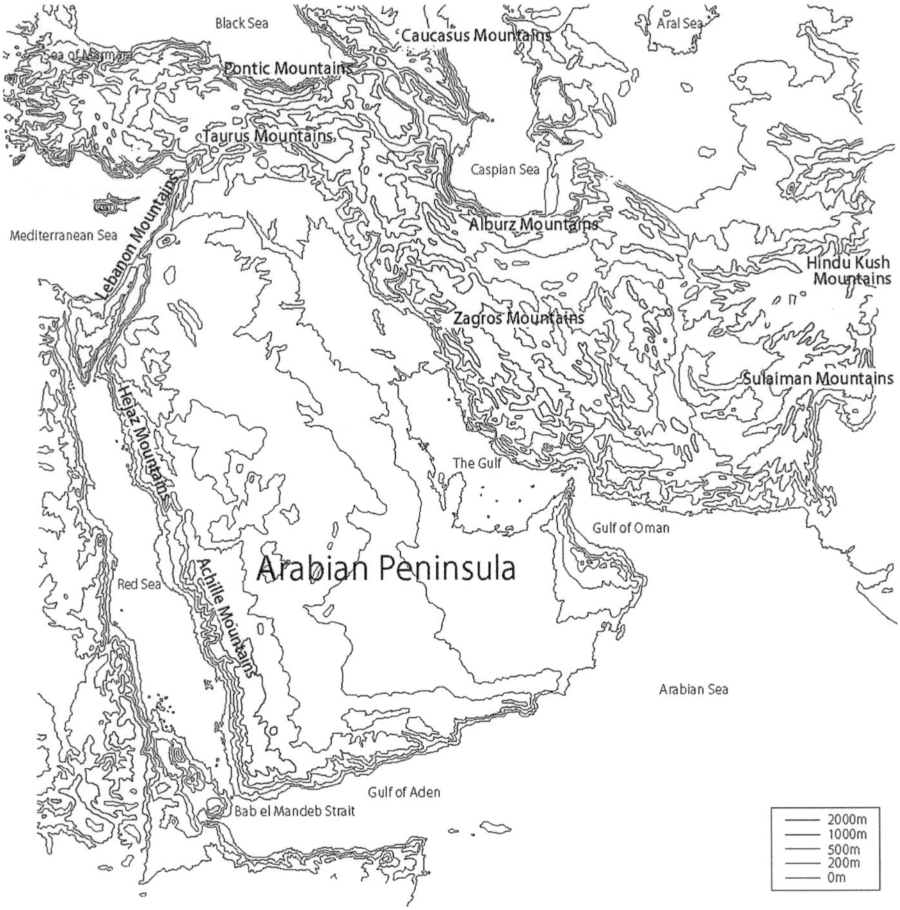


**Fig. 2.2** Large topography of West Asia

civilization prospered in the area where the Tigris and Euphrates basins reach the Southern Levant. This semicircular area is known as the Fertile Crescent (Fig. 2.3) in reference to its rich agriculture compared with the neighboring barren desert zones.

Although most parts of the Arabian Peninsula are desert zones (Fig. 2.3), including the Rub al Khali and Nafud deserts, the area facing the Red Sea is a high mountain belt more than 2000 m above sea level (Fig. 2.2). Volcanoes are distributed in this high mountain belt and are concentrated near the Bab-el-Mandeb strait. Parts of these volcanoes are said to constitute the Great Rift Valley, which includes the famous lava lake of Erta Ale volcano (Fig. 2.4). The Great Rift Valley, which is now inactive, extends to the south from the northeastern part of the African Continent to South Africa, separating East Africa from the rest of the continent. Its eastern exten-

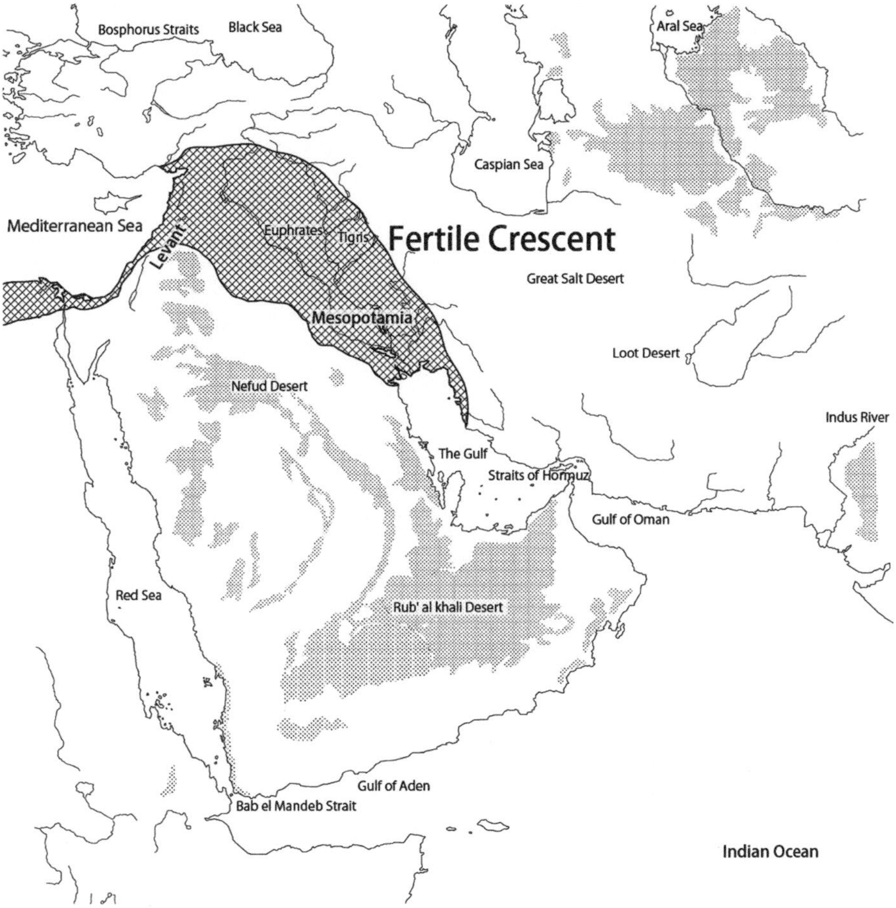


**Fig. 2.3** Deserts and large revers of West Asia

sion forms active spreading oceanic ridges in the central Indian Ocean. Most of the oldest human remains were discovered in the Great Rift Valley.

### 2.3.2 *The Origin of the Desert*

In this section, the origin of the desert will be considered, which characterizes the nature of West Asia. Desert is defined as land in which the evaporation rate is greater than the precipitation level. There are three types of deserts: (a) those occurring near the equator within latitudes of 20°–30°, (b) those formed near the coast by offshore cold currents, (c) and those formed inland behind great mountain ranges.

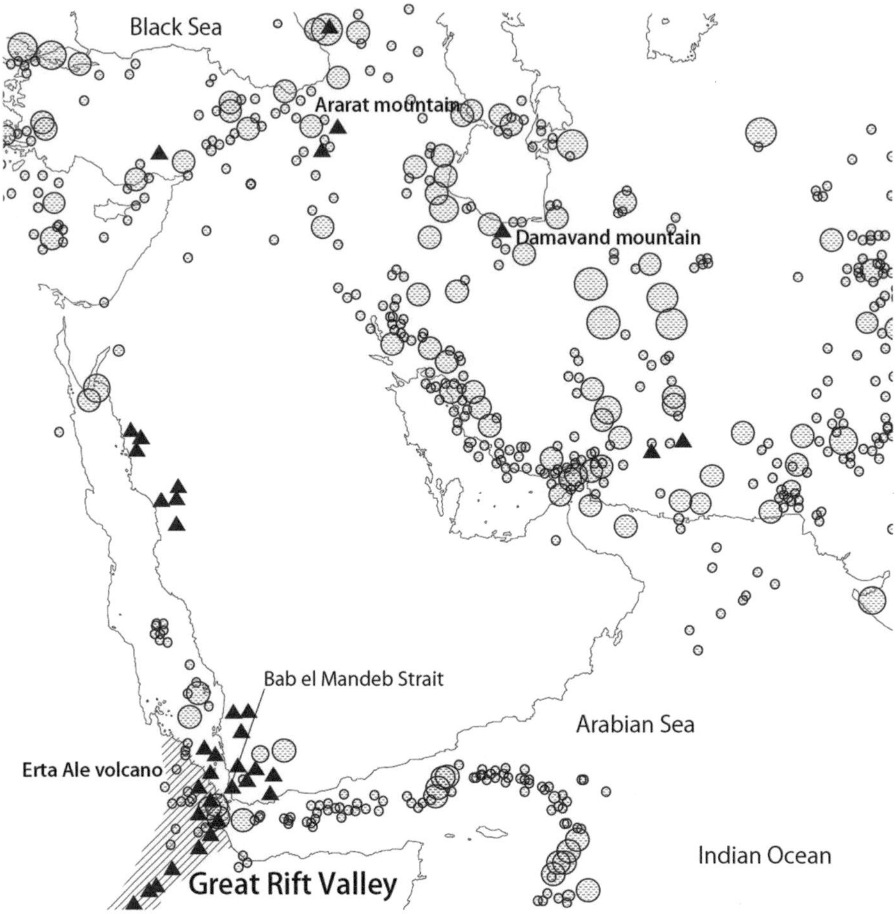


**Fig. 2.4** Active volcanoes (*triangles*) and earthquake epicenters (*circles*) of West Asia. The circle size indicates earthquake magnitude

Type (a) desert is established in the subtropical high pressure belt where dry air of high temperature descends. In the case of no rotation of the Earth, air warmed in tropical zones rises and descends in the Polar region. This phenomenon is known as Hadley circulation. However, strong air currents flowing in the north and south directions caused by Earth's rotation create a whirlpool effect. As a result, Hadley circulation occurs only in zones of low latitude where the influence of the rotation is very small. In zones of high latitude, however, where the influence of the rotation is large, Rossby circulation occurs. Atmospheric circulations in low and high latitudes are referred to as Ferrel and Polar cells, respectively. A subtropical high pressure belt lies between the border of the Hadley–Rossby circulation at about 20°–30° in latitude. Clouds do not appear over the land in this subtropical high pressure belt, and the precipitation is very small. This condition has created deserts such as the Sahara and Rub-al Khaliin the Arabian Peninsula.

**Fig. 2.5** Wadi in the Zagros Mountains

Type (b) desert is produced by cold sea currents that flow from the polar regions toward the equator. Large quantities of heat occurring on land are cooled when contacting the cold sea currents. Because this cool and heavy air does not ascend, it does not form a rain cloud. Instead, it remains inland and forms deserts such as the Atacama and Namib.

In type (c) desert, air containing moisture meets and collides against high mountains along the shore, and precipitatesrain and snow. The remaining dry air across the mountains may accompany abnormal temperature rise to create desert conditions. This mechanism, known as the Föhn phenomenon, occurs unexpectedly and results in abnormally high temperatures. The Takla Makan desert was developed by this phenomenon.

Oasis, a freshwater catchment in a dry zone with a groundwater spring originating from rivers or melt water may be distributed in such desert. Because of the high surface runoff rate due to dry, barren, desert landscape, seasonal rivers, known as *wadi* in Arabic (Fig. 2.5) are common feature in the West Asian landscape, where water flows only during temporal heavy rainfall events.

**Fig. 2.6** Limestone cave in Zagros Mountains

### 2.3.3 Limestone Caves

A widespread limestone field stretches from the Zagros Mountains to the Mediterranean Sea coast. The limestone caves in this field played an important role during the Paleolithic Period as human dwellings. Limestone is easily dissolved in rain waters that is saturated by atmospheric carbon dioxide and typically have a pH ~ 5.6. As a result, the limestone fields develop distinctive landscapes. "Karst", the term originates from the Karst District of Slovenia, is characteristic landscape of limestone belts and includes topographic features such as dolines, karrenfields, and limestone caves (Fig. 2.6). A doline is a funnel-shaped depression formed on the ground surface of a limestone zone. Rainwater soaks into a doline and flows into the basement to become groundwater. Pillars of white limestone formed by dissolution action with the rainwater occur around a doline. The topography with the surface unevenness is referred to as karren; land with karrens is known as a karrenfield.

Active limestone caves usually involve development of stalactites, which hang from the ceiling, and stalagmites, which are shaped like mushrooms on the floor. The dissolution of a limestone cave begins with groundwater flowing along cracks in a saturated water zone beneath the water table. The rain on the ground surface reaches the saturated water zone through underground cracks. The region between the ground surface and the saturation zone is known as a circulation water zone. The water flow channels migrate horizontally and vertically, and the duct of the saturated water zone gradually spreads laterally. Limestone is dissolved mostly in the shallow part of the saturated water zone in a semicircular shape, and stalactites are formed in the upper part of the cave space. When the water table drops, the dissolution finally stops, and the cave enters a stable stage.

As exemplified by the active Akiyoshido Cave in Japan, water generally plays a crucial role in developing the limestone caves. In the case of the Akiyoshido cave, there are three development stages according to the location level: high terrace (300 m asl), middle terrace (200 m asl), and low terrace (100 m asl). This situation indicates that the saturation water zone has decreased gradually. The caves on each terrace seem to have been formed 500,000–1 million years ago, 200,000–500,000 years ago, and 100,000–200,000 years ago, respectively.

In contrast to the East Asian Monsoon climate, the precipitation in West Asia is considerably lower. The monthly precipitation from November to April is between 20 mm and 80 mm in Shiraz, southern Iran. The precipitation is extremely low in this region, at 0 mm–7 mm from May to October. The formation of a limestone cave under such climate conditions seems impossible. However, the formation processes of limestone caves continue over tens of thousands of years, as exemplified by the Akiyoshido Cave. The limestone caves in southern Iran must also have been formed over thousand of years during which climatic conditions may have been quite different from those of the present day.

## 2.4 Geological Features of West Asia

### *2.4.1 General View*

When looking at the geological features of West Asia, we immediately notice the differences in features across a line connecting the Gulf and the Mediterranean through the Fertile Crescent (Fig. 2.7). Large-scale distribution of Precambrian rock occurs on the south side, whereas a fold belt has developed on the north side. Fragments of oceanic crust known as ophiolites are distributed in this fold belt, which extends to Oman, an edge of the Arabian Peninsula. The largest ophiolite

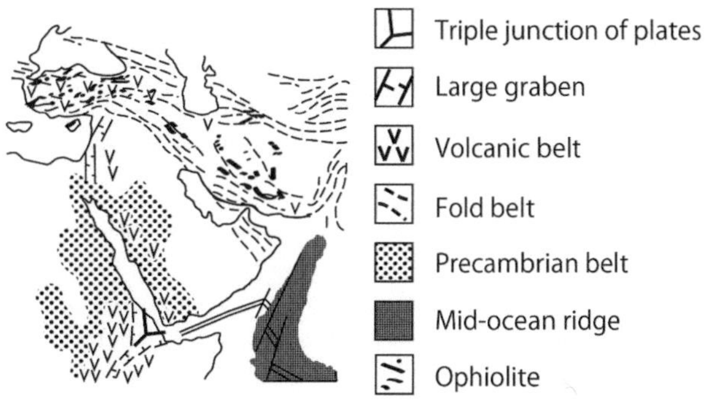


**Fig. 2.7** Geologic outline of West Asia

block ever known is exposed in Oman referred to as Oman ophiolite. The expanse of the Precambrian rock suggests that the Arabian Peninsula was a part of a stable landmass, the African Continent. When a continent breaks, it usually breaks forming three rift zones radially disposed about a triple junction. Then, two of the three will evolve to matured rift zones or spreading centers, whereas the other usually becomes a failed rift known as aulacogen. Active spreading ridges in the Red Sea and the Gulf of Aden, and the East Africa Great Rift Valley, that is currently inactive, were developed in such a manner. In addition, volcanoes are usually distributed at triple junctions and along rift valley extension. The formation of such volcanic fields characterizes a plate divergent zone. In the fold belt, carbonate rocks and clastic rocks, which are accumulated in the Tethys Sea, are widely distributed hosting the ophiolites. Unlike the current Pacific Ocean, it is thought that the Tethys was generally a shallow sea.

Sedimentary rocks, which are formed through consolidation of sediments, can be classified into two types. The first category is terrigenous clastics, which are formed by transportation and accumulation of terrigenous grains made of weathered and eroded surface rocks. The second includes carbonates, siliceous rocks, and chemical precipitates, which are formed by organisms or chemical deposition. Carbonate rocks which are the most representative rocks in West Asia will be described first. A discussion of the archaeologically important siliceous rock will follow it.

## *2.4.2 Carbonate Rock*

### 2.4.2.1 Coral and Zooxanthellae

Carbonate rocks are considered to form through organic processes: some carbonates are directly related to structures such as coral reefs. Coral reefs are made of skeletons of coral polyp, a member of Coelenterata. The coral polyps grow by eating plankton, and develop their skeletons using calcium and bicarbonate ions dissolved in marine water. In addition, the coral polyps receive nourishment by photosynthesis from zooxanthellae living inside their tissue. Therefore, coral polyps may live as deep as they can photosynthesize; such a depth is known as the coral growth depth. The growth of coral depends on water temperature, salinity, photoenvironment, waves, and water flow.

### 2.4.2.2 Formation of Carbonate Rock Bodies

Carbonate rock is a generic term for sedimentary rocks with a ratio of carbonate minerals such as a calcite, aragonite and dolomite of more than half the volume. Carbonates are compounds in which part or all of the hydrogen of carbonic acid ($H_2CO_3$), a weak acid generated when carbon dioxide is dissolved in water, is replaced by metal.

**Fig. 2.8** Carbonate platform of Dalnesin Range, southern Iran. The difference in elevation is more than 1000 m

The formation of carbonate sediments, which become carbonate rock through lithfication, depends on (1) organic production; (2) supply of terrigenous sediments; and (3) erosion, transportation, and sedimentation (Hoyanagi et al. 2004). Carbonate is produced by organisms in coral reefs, and the supply of the terrigenous sediments determines the production of the carbonate sediment. Carbonate minerals are also produced in subarctic and temperate climatic zones where the supply of the terrigenous clastics is poor.

Because particles constituting carbonate sediments are produced by organisms, its formation is affected by factors such as water temperature, light quantity, salinity, sedimentation rate, and the water circulation.

Reef corals are the main constituents of carbonate sediments in the tropic zone, whereas benthic foraminifera, bryozoans, and bivalves are the main producers in temperate and subarctic zones. However, reef coral is overwhelmingly predominant in the origin of carbonate mineral particles.

Carbonate sediments often form hills and plateaus, and the body of carbonate sediments is known as a carbonate platform (Fig. 2.8). These carbonate platforms are classified into the following categories:

(a) Rimmed continental shelf: This type occurs when a reef or the barrier of the carbonate sand body is formed on a remote continental shelf far from land. The Florida Peninsula and the Great Barrier Reef belong to this type.

(b) Carbonate ramp: This type is formed in a continental shelf having a gentle slope, and it resembles a rimmed continental shelf. However, farther from land and at deeper depths, it does not form a reef or barrier on the continental shelf. The Gulfbelong to this type.
(c) Epeiric platform: Different in size from the previous two types of carbonate platforms, this type extends hundreds and thousands of kilometers and was formed near the Gondwana Continent in the early Paleozoic. No example of this type exists in modern times.
(d) Isolate platform: This type stands alone in the deep sea surrounded by straits. A coral reef develops in the periphery area, and its central part tends to be a depression or lagoon that forms an atoll. The Bahama Banks are a famous example, and seamounts and crest areas of fault scarp of the eastern Red Sea margin belong to this type.

The representative carbonate rock occurringon land is limestone. This rock is composed of brecciated sediment consisting of coral reef material formed in a shallow sea less than several hundred meters in depth.

### *2.4.3 Siliceous Rocks*

Siliceous rocks are rich in silica ($SiO_2$) and were very important for making lithic implements used by humans. Cenozoic siliceous shale and the Mesozoic/Paleozoic bedded chert are typical siliceous rocks occurring in the Japanese Islands.

Chert contains more than 90 % silica in the form of cryptocrystalline and microcrystalline quartz. It contains chert layers several centimeters to ten and several centimeters in thickness composed of siliceous remains such as radiolarian tests, which alternate with layers of mud several millimeters to tens of millimeters thick that do not contain abundant siliceous remains.

The accumulation state of organic remains closely resembles siliceous mud, which is deposited in deep ocean floor. The sedimentation rate is estimated to be approximately 1 mm per 1000 years based on radiolarian biostratigraphy. Together with the absence of sand grains the bedded chert is considered to bepelagic sediment. However, sediment equivalent to such a stratified chert formation has not been discovered in the modern ocean floor. The origin of the bedded sediment is suggested as follows:

(a) Periodical fluctuation of siliceous plankton
(b) Secondary deposition of siliceous sediment (turbidite)
(c) Chemical segregation

Of these, (a) explains the chert layer formed during blooming and rapid accumulation of radiolarians, whereas the mud layer formed during the period of low radiolarian production. (b) explains deposition from turbiditic flow sorted particle size into coarse radiolarian tests and fine clay particles. (c) attributes the siliceous layer

**Fig. 2.9** Radiolarite (*right*; from radiolarite outcrop, *left*; from excavation site)

to chemical accumulation of silica dissolved from mixed sediments of radiolarian tests and mud during diagenesis.

In addition, chert occurs in a nodular form, which archaeologists refer to as flint. It occurs as ball-like and irregular-shaped chert in carbonate rock, and it consists of microcrystalline quartz. This silica formed from radiolarian tests and sponge spicules and was crystallized as a result of a kind of segregation.

In addition to these types of chert, radiolarian rock is also common in West Asia (Fig. 2.9). Radiolarite is a homogeneous muddy sedimentary rock containing a high density of radiolarian fossils. This type is not as hard as bedded chert. Radiolarian strata are often accompanied by limestone formation as acomponent of the Zagros fold belts in south Iran. According to microscopic observation of radiolarite, it is characterized by the existence of foraminifera tests scattered among radiolarian tests.

### *2.4.4 Ophiolite*

Ophiolites are believed to be fragments of oceanic lithosphere, several kilometers to tens and several kilometers thick, that were obducted to the neighboring continental crusts. An ophiolite usually consists of, from above, basalts (extrusive volcanic rocks), dolerite dikes, gabbros (mafic crystalline rocks) and ultramafic rocks (peridotites and serpentinites) (Fig. 2.10). Pelagic chert or terrigenous sediments may cover the oceanic igneous rocks. Among the ophiolite constituents, basaltis the igneous rock with the widest distribution on Earth surface because it constitutes the

**Fig. 2.10** Peridotite outcrop in Neyriz ophiolite, southern Iran

ocean floor, and large igneous provincessuch as flood basalts in the Deccan trap. An outline of ophiolites, which frequently occur in the fold belt in West Asia, is explained here.

#### 2.4.4.1 From the Formation of Basaltic Magma to Its Extrusion

In oceanic environments, basaltic magmas are usually generated underneath mid-oceanic ridges by partial melting of peridotites in the upper mantle. The melt separates from the partially moltenperidotite and accumulates to form magmas. The remaining rock after melt extraction is known as residual peridotite. When magmasform a magma reservoir in the crust, it begins to cool to form new oceanic crust. During the cooling, crystals that formed through crystallization in the magma chamber sink through it and accumulate at the bottom due to the density contrast between lighter magmas and heavier crystals. Rocks produced by this crystal settling process are known as cumulates. The rest of the magmas may continue to rise, and finally reach to the earth's surface. Then, it is cooled rapidly by marine water and form pillow lavas of basaltic compositions.

#### 2.4.4.2 Question About the Formation of Ophiolites

A large number of isolated ophiolite blocks are distributed over the fold belts in West Asia. It was believed that ophiolites represent structures of the oceanic lithospheres. However, as research on world ophiolites progressed, the chemical composition of its volcanic rocks were revealed to often have a similar composition to those formed in a subduction zone environment. Therefore, most ophiolites in West Asia are believed to have been developed in the so-called supra-subduction zone environment.

Ophiolites often accompany low-angle fault planes at its lowest boundary. When a continental landmass led by an oceanic plate reaches the subduction zone, the landmass cannot sink owing to its buoyancy. As a result, it is thought that the tip of a mantle wedge in the subduction zone thrust onto the landmass in a process known as obduction. It is often known that the formation age of igneous rocks in ophiolites and the age of arc magmatism intruded into the ophiolite are similar, which suggests that the formation sites of ophiolites are near island arcs or marginal seas where igneous activity was high (Arai 1988). Accordingly, because abundant ophiolites occur in the West Asia fold belt, numerous obduction events must have occurred in this region due to multiple crustal movements throughout the geological timescale.

### *2.4.5 History of Geological Feature in West Asia*

During middle Paleozoic (3–4 hundred million years ago) to early Cenozoic (several tens of million years ago), the Tethys Ocean extended from east to west along the equator. During a part of the period, from Permian (the latest Paleozoic) to Triassic (early Mesozoic), Pangaea, a vast supercontinent comprising all the continents of the earth, appeared. Pangaea was situated astride both hemispheres; the Northern Hemisphere section was known as Laurasia, and the Southern Hemisphere side was known as Gondwana. Meanwhile, the Tethys opened toward the east in a fan shape with a hinge near the present-day Mediterranean Sea area. There were two oceans at that time. The first was the Tethys, serving as a huge inlet of Pangaea, and the second was Panthalassa, or the old Pacific, which surrounded Pangaea and the inlet from the outside (Fig. 2.11a).

Although Pangaea was a single huge continent, it underwent repeated breakup and collage. Tectonics in Pangaea are characterized by breakup along the northern edge of Gondwana and collage of small continents along the southern edge of Laurasia during several hundred million years. A typical example is the divergence and the convergence of Cimmerian Continent.

The Cimmerian Continent was a strip-shaped continent that was a collage of small continents from current West Asia to current Southeast Asia. The present-day Turkey, Sanandaj–Sirjan in the central part of Iran, south Tibet, and Sibumasu ter-

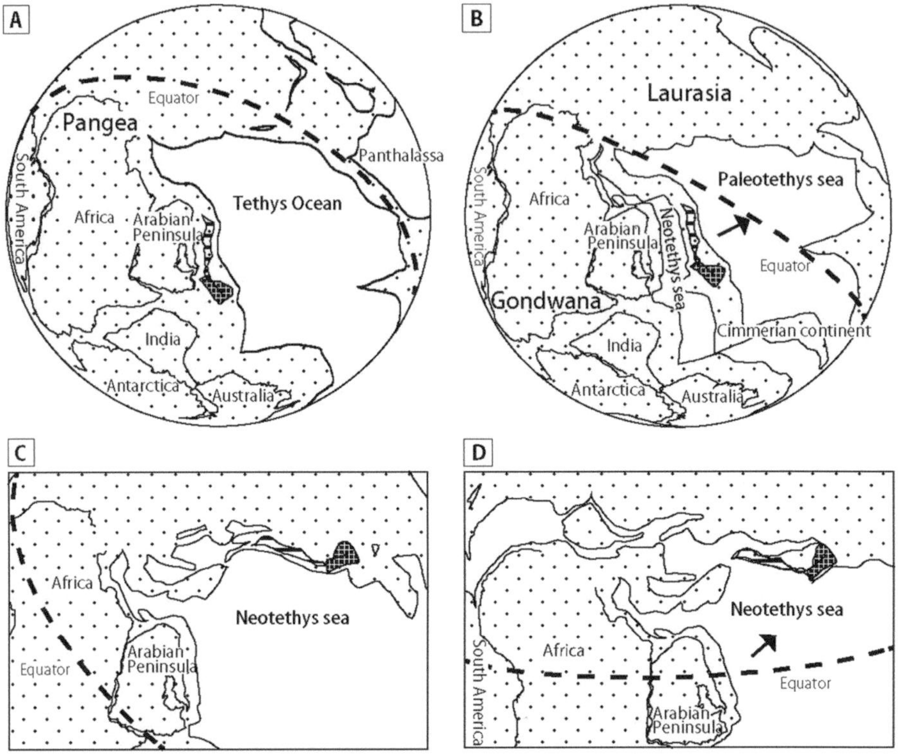


**Fig. 2.11** Paleogeographic maps of West Asia (**a**); Carboniferous period of Paleozoic era, (**b**); Permian period of Paleozoic era, (**c**); Jurassic period of Mesozoic era, (**d**); Cretaceous period of Mesozoic era. After Cavazza et al. (2004)

rane extending from Myanmar to Sumatra through Thailand and Malaysia, were aligned to form this continent (Fig. 2.11b). This strip-shaped continent diverged from Gondwana, which produced a new ocean that separated from the old Tethys. This new ocean is known as the Neotethys, whereas the old ocean is known as the Paleotethys, to distinguish from the former Tethys Ocean. The expansion of the Neotethys was caused by the drift of the Cimmerian Continent to the north and accompanied consumption of the Paleotethys. The reduction and eventually the disappearance of the Paleotethys was caused by the subduction of the Paleotethys plate into the trench in front of Laurasia. Ultimately, the Cimmerian Continent combined with Laurasia during Triassic to Jurassic; the Alborz and other mountains were formed by this activity (Fig. 2.11c). In the Cenozoic era, the Arabian plate diverged from the African plate; thereafter, the Arabian plate drifted to the north and collided with the Eurasian (Iranian) plate. This collision formed the Zagros Mountains (Fig. 2.11d).

## 2.5 Richness of Mineral Deposits in West Asia

### *2.5.1 Petroleum*

To say that West Asia is supported by abundant petroleum is not an exaggeration. Modern industry depends on petroleum; thus, West Asia has strongly affected the world's economy. In this section, the reasons for the formation of such a huge petroleum field in West Asia are examined.

As for the origin of the petroleum, "origin hypothesis of Kerogen" is convincing. Kerogen is produced by high molecular compounds derived from organisms such as plankton, which are altered by terrestrial heat and pressure in a lengthy process known as pyrolysis. Rock containing more than 1 % gross weight of kerogen is known as petroleum source rock.

When we consider that the kerogen is derived from plankton, it appears that the sedimentation environment was the reductive with fine-grained sediments. Petroleum source rock is formed deep in a closed ocean basin such as the Black Sea. About half of the world's petroleum source rock is the strata accumulated between the late Jurassic and the early Cretaceous eras. During this period, the temperature was highest in Earth's history. The polar regions lacked ice, and oceanic general circulation might not have occurred. Therefore, the world ocean floor was reductive, which accelerated the formation of petroleum source rock (JOGMEC 2010).

The petroleum soaks into rocks with high porosity known as petroleum reservoir rock; 56 % of petroleum reservoir rocks in the world's huge petroleum field are sandstones, and 44 % are carbonate rocks. Sandstone is composed of completely spherical grains, which represents the closest packing, has a porosity of 26 %. Porosity of 20 % and 15 % are considered to be very superior and good petroleum reservoir rocks, respectively; less than 5 % porosity is not considered a petroleum reservoir rock. The petroleum is produced in petroleum source rock and moves into the petroleum reservoir rock with high porosity during its primary movement. In its secondary movement, the petroleum is transmitted through the reservoir rock and moves to upper underground regions. The petroleum finally is stored in underground tanks known as traps, which form a petroleum field. Impermeable rock is necessary for forming these tanks because the petroleum must remain in place underground. This rock is also known as cap rock and is often composed of mudstone or halite. The buoyancy of petroleum results in high pressure applied to the upper level of the cap rock, and in some cases, petroleum penetrates the cap rock and seeps onto the ground surface. It is well-known that the ancient people of Baku near the Caspian Sea and the Tigris basin, Iraq, used such seeped petroleum for fuel and ointment (JOGMEC 2010).

As previously mentioned, the elements necessary for establishing West Asia as a petroleum field include petroleum source rock containing abundant kerogen, reservoir rocks such as sandstone and carbonate rock, and suitable tectonic movement

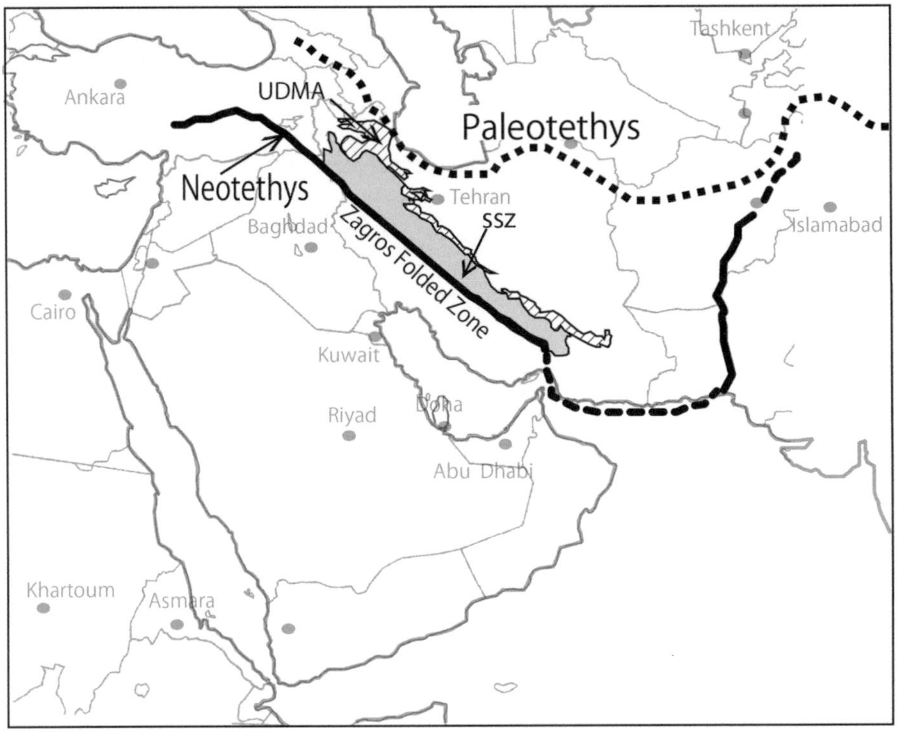


Fig. 2.12 Paleotethys and Neotethys (Aghanabati and Ghorbani 2011) *SSZ*, Sanandaj-Sirjan Zone, *UDMA*, Urumiyeh-Dokhtar Magmatic Arc

such as a fold belt. In addition, the ex-equator belt provides a suitable zone because of the huge accumulation of petroleum reservoir rock and carbonate rock. Therefore, the reasons for the concentration of the large petroleum field in West Asia is explained by 1) the existence of the Zagros fold belt, 2) carbonate rock and sandstone sedimentation in the Tethys Ocean, and 3) abundant kerogen in the petroleum reservoir rock.

## 2.5.2 Metallic Minerals

In addition to petroleum, West Asia is a region of rich metallic minerals, particularly those useful for humans.

The tectonic divisions around the Zagros Mountains are as follows from north to south are: the Cimmerian Continent including the Alborz Mountains, Central Iran, the Urumiyeh–Dokhtar Magmatic Arc (UDMA), the Sanandaj–Sirjan Zone (SSZ), and the Zagros Folded Zone including the Zagros Mountains (Fig. 2.12). The northern and southern parts of Turkey are known as the Pontides and Anatolides–Taurides, respectively. The latter belongs to the Cimmerian Continent. The UDMA, which is situated between the SSZ and Central Iran, runs parallel to the Zagros and the SSZ. This arc bears huge volcano-sedimentary deposits more than 10 km thick in some places and forms a topographic ridge separating the SSZ from Central Iran. The UDMA is characterized by Paleogene magmatism, predominantly arc or island-arc type. Volumetrically, the volcanic rocks were produced mostly during the Eocene (Agard et al. 2011). The SSZ forms the continental collage of Iran together with the Lut block and other blocks from Central Iran and extends from the Bitlis area in Turkey to the western end of Makran. The crustal root of the Zagros orogeny coincides with the SSZ with a crustal thickness of 55 km–70 km. The metamorphosed and deformed SSZ is also characterized by the emplacement of subduction-related, mainly Mesozoic calc–alkaline plutons and lavas. Age compilation of recent radiometric dating has confirmed that the SSZ is the locus of arc magmatism that occurred mainly during the Mesozoic, although recent data shows local magmatic activity as young as the Eocene in the northwest SSZ (Agard et al. 2011).

The Cimmerian Continent zone includes back-arc, magmatic arc, fore-arc, trench, ocean floor, and likely mid-ocean ridge tectonic settings formed in the Mesozoic that amalgamated through collision processes. This resulted in the formation of metallogenic zones along the orogenic belts. Jankovic and Petrascheck (1987) distinguished five groups of metallogenic zones and districts in the Alpine–Himalayan Belt related to late orogenic activation, subduction and collision, ocean floor spreading, rifting, and Alpidic remobilization.

In terms of plate tectonics, the locations of certain kinds of mineral deposits can be expected (Skinner et al. 2013). In a magmatic arc, copper, gold, silver, tin, lead, mercury, and molybdenum from copper porphyry deposits are expected to occur in veins. In a back-arc basin, copper, zinc, gold, and chromium from volcanogenic massive sulfide occur in stratabound deposits and evaporates. In West Asia, these deposits are mining targets that are categorized according to type of magmatic arc and back-arc basin (Figs. 2.13 and 2.14). In addition, iron deposits are variously formed as sedimentary mineral deposits, magmatic mineral deposits, and hydrothermal mineral deposits, although most iron deposits are produced from sedimentary mineral deposits (e.g., Lake Superior-type iron deposits). Iron deposits derived from magmatic mineral deposits and hydrothermal mineral deposits are also conspicuous (Fig. 2.15).

As shown in Figs. 2.13, 2.14 and 2.15, metal deposition sites are concentrated in the UDMA and SSZ. Therefore, districts north of the Zagros Mountains are characteristically rich in metals compared with southern districts except for the Red Sea area.

### *2.5.3 Raw Materials for Stone Tools*

Siliceous rocks were also important for the ancient people in West Asia. The raw materials used to make chipped stone implements included chert, obsidian, siliceous nodules, radiolarite, and other siliceous rock. Two famous obsidian sources

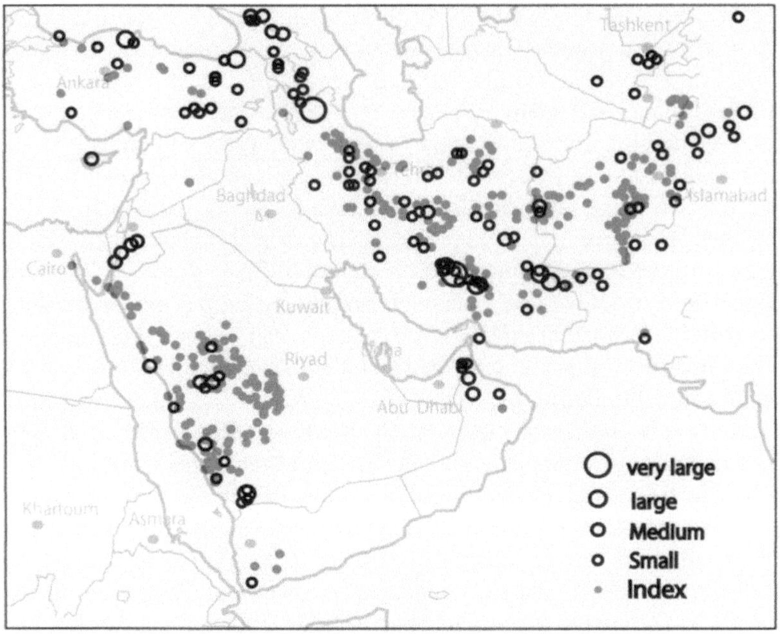


**Fig. 2.13** Copper deposits (Aghanabati and Ghorbani 2011)

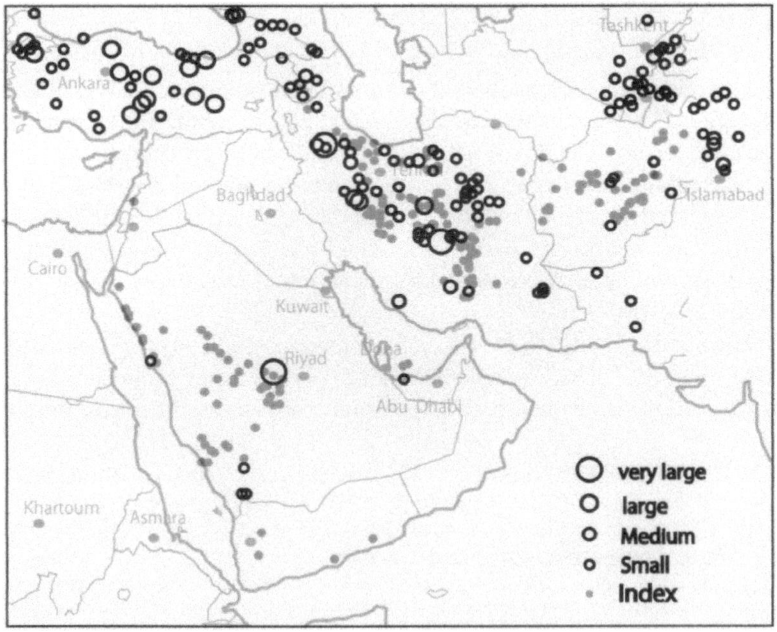


**Fig. 2.14** Lead-zinc deposits (Aghanabati and Ghorbani 2011)

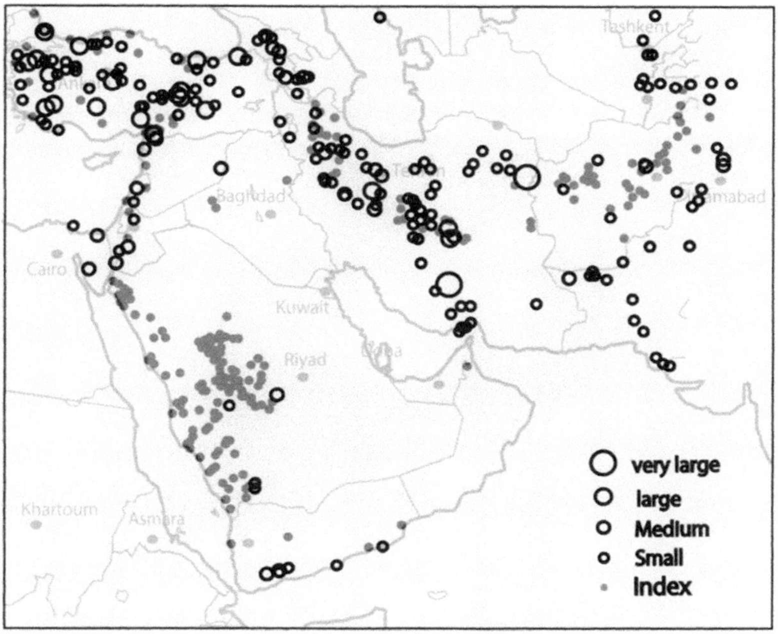


**Fig. 2.15** Iron and manganese deposits (Aghanabati and Ghorbani 2011)

are situated in Anatolia, and volcanic glass which has been traded throughout West Asia was used for making chipped implements, particularly during the Neolithic era. Other siliceous rocks such as chert and radiolarite have wider distributions and were indispensable materials for making chipped stone implements. I personally had a chance to study radiolarite of the Iranian Zagros Mountains, and have previously reported the general characteristics of siliceous rocks. Therefore, the present discussion is focused on radiolarite, particularly that from the Zagros region. Radiolarite has several advantages as a raw material for chipped stone implements such as homogeneity, hardness, and the capacity for splitting. These features are all crucial for making stone tools.

Radiolarite is the comparatively hard fine-grained chert-like homogeneous consolidated equivalent of radiolarian earth. Although radiolarite and chert are collectively classified as siliceous rocks, their degree of crystallinity is generally different. Chert is a microcrystalline or cryptocrystalline sedimentary rock consisting predominantly of interlocking crystals of quartz less than about 30 μm in diameter. Although chert presents a more glassy appearance and creates a sharper edge than radiolarite, it is more fragile. A striking feature revealed under a microscope is that radiolarite consists of radiolaria as well as planktonic foraminifera (Fig. 2.16). Thus, the presence of planktonic foraminifera in radiolarite means that the radiolarite was deposited above the calcite compensation depth (CCD). In the ocean, the rate of solution of calcium carbonate exceeds the rate of its deposition below the CCD; thus, foraminifera shells cannot be present below the CCD. In the Pacific

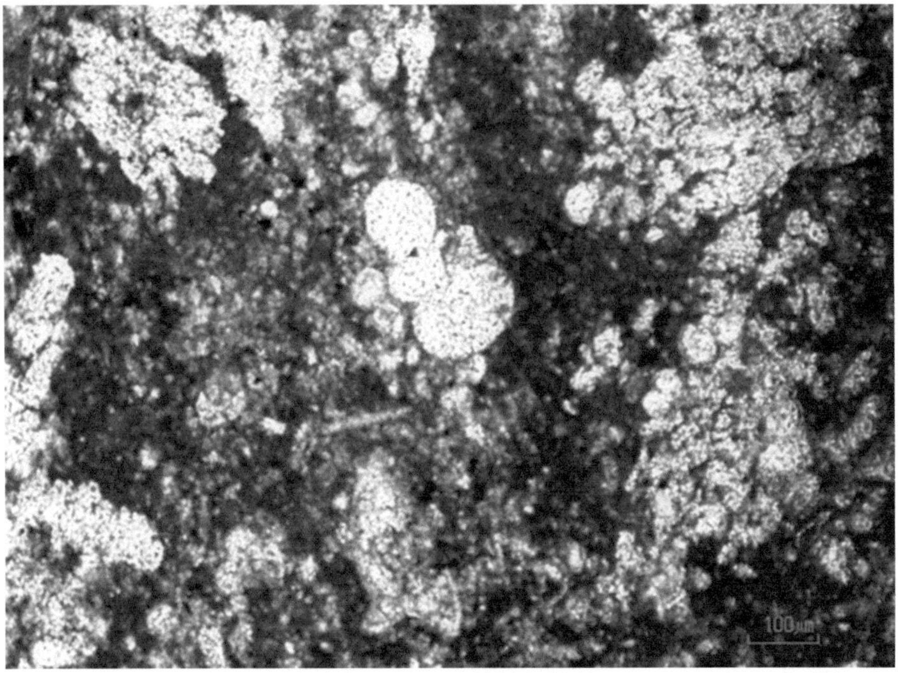


**Fig. 2.16** Planktonic foraminifera in radiolarite from near Arsanjan, southern Iran (Identified by Dr. H. Uematsu)

Ocean, this level is at 4000 m–5000 m; the level in the Atlantic is somewhat shallower. It is inferred that the depth of the sea bottom in which the radiolarite of the present-day Zagros Mountains was deposited was shallower.

The question remains of how deep sediment such radiolarite can coexist with shallower rock such as limestones. The answer can be found in the geodynamics of the Zagros Mountains. The Zagros orogeny fundamentally reflects the collision tectonics between the Arabian plate and the Central Iran microcontinent that occurred during the Paleogene Period. The components of the Zagros Crushed Zone, SSZ, and UDMA were located between the plate and the microcontinent, and through collision tectonics, the components of these three zones were thrust over the Zagros Folded Zone on the Arabian plate. The Zagros Crushed Zone includes an ophiolite suite, which is an association of ultramafic rocks, coarse-grained gabbro, coarse-grained diabase, volcanic rocks, chert, and radiolarite. In short, the deeper radiolarite was thrust over the shallower limestone.

Biglari (2004) reported that radiolarite was used in present-day Kermanshah as a raw material for stone tools. Radiolarite occurs in the Zagros Mountains longitudinally and intermittently. Ancient people utilized radiolarite for stone implements and limestone caves for residences soon after arriving in the Zagros Mountains. They may have acquired skills and improved their technologies for making stone implements.

## 2.6 Conclusions

West Asia is known as the birthplace of ancient civilization, and the fertile nature of the locality has attracted attention. In addition, the huge petroleum region that became the base of modern industry spreads throughout West Asia with good reserves.

*Homo sapiens* left Africa and settled into West Asia 160,000–80,000 years ago (Kadowaki 2015). Thereafter, they achieved great human transformation such as settled habitation, neolithization, and urbanization. For these innovations, the geology of West Asia must have played indispensable roles in each era. The main points covered in this paper are summarized below.

1. Concerning the plate tectonics, West Asia was formed as a node of three continents. This formation caused the geological complexities of West Asiawith the concentrations of stone and mineral resources along the plate boundaries, which were so significant for the development of lithic culture and metal use in human societies.
2. In the Paleolithic West Asia, the geology surrounding the Fertile Crescent provided an extremely attractive environment for humans including large limestone caves for their habitation and wonderful siliceous rocks such as chert and radiolarite for making chipped stone implements.
3. In the Neolithic era, species of plants and animals were domesticated. In addition, good geologic materials were available for building houses, such as limestone and basalt, and suitable siliceous rocks for making agricultural tools such as sickle elements. This also demonstrates that the geology of West Asia was extremely attractive for providing the basic requirements of humans.
4. After the Bronze Age, basic metallic materials such as copper, tin, and iron that were used to produce highly advanced tools existed in particular areas of West Asia. No other area in the world offers such metals in combination. These geologic requirements are prominent among areas of ancient civilization. It cannot be overemphasized that civilization occurred in West Asia as a result of these geologic requirements.
5. When we consider the role of petroleum in modern world, it is quite clear that petroleum has a special position in the world politics and economy. Siliceous rocks, metallic minerals including iron, and other important precious stones and their combinations had a special position in each era from the Paleolithic to the present world. This demonstrates that geological features became the basis of culture. Therefore, I answer the title of this paper with "Yes" without hesitation.

## References

Agard, P., J. Omani, H. Jolivet, H. Whitechurch, B. Vrielynck, W. Sparkman, P. Monie, B. Meyer, and R. Wrtel. 2011. Zagros orogeny: A subduction-dominated process. *Journal of Geological Magazine* 148(5–6): 692–725.

Aghanabati, A., and Ghorbani, M. 2011. *Metallogenic map of the Middle East, 1:5,000,000*. Geological Survey of Iran. Iran: Tehran.
Arai, S. 1988. Mantle overthrusting onto earth's surface-crustal slice- ophiolite. *Kagaku (Science)* 58(11): 685–695. (in Japanese).
Biglari, F. 2004. The preliminary observations on middle palaeolithic raw material procurement and usage in the Kermanshah Plain, the case of Do-Ashkaft Cave. In *Persian Antiques Splendor, mining crafts and archeology in ancient Iran*, vol. 1, ed. T. Stollner, R. Stotta, and A. Vantadoust, 130–138., Deutsches Bergbau-Museum Bochum.
Cavazza, M., Roure, F.M., Spakman, W., Stampfli, G.M., and P.A. Ziegler. eds. 2004. *The Transmed Atlas*, 141. Berlin/Heidelberg: Springer.
Hoyanagi, K., Kumon, F., and H. Matsuda. 2004. *Sediments and sedimentary rocks*, 171. Tokyo: Kyoritsu Shuppan. (in Japanese)
Jankovic, S., and W.E. Petrascheck. 1987. Tectonics and metallogeny of the Alpine-Himalayan belt in the Mediterranean area and Western Asia. *Episodes* 10(3): 169–175.
JOGMEC (ed.). 2010. *Whereabouts of oil resources-how much of oil resources is remaining*, 188. Tokyo: Corona Publishing. (in Japanese)
Kadowaki, S. 2015. The distribution pattern of archaeological objectsfollowing the diffusion of *Homo Sapiens* in North Africa, West Asia andEurope. In *Homo Sapiens* and the Neanderthal, ed. Y. Nishiaki, Vol.3, 3–19. Tokyo: Rokuichishobo. (in Japanese)
Skinner, B.J., S.C. Porter, and J. Park. 2013. *Dynamic earth*, 584. New York: Wiley.

# Chapter 3
# A Brief Review of the Geology of West Asia

**Seyed A. Aghanabati**

This manuscript offers a brief review of the geological and cultural history of the western part of Asia (hereafter, WA), which includes Azerbaijan, Afghanistan, Abu Dhabi, the Caspian Sea, Iran, Iraq, Oman, Pakistan, Qatar, Turkmenistan, Turkey, and the United Arab Emirates. There is a particular emphasis on the comprehensive relationship between the earth sciences and the social and cultural aspects of WA, parts of which are also thought of as being part of the Middle East.

## 3.1 Topography and People

Varied climatic and topographic conditions prevail over the vast expanse of WA, which extends between 8–42° N latitude and 24–88° E longitude. Deserts are a common feature, including those in the Arabian Peninsula, Iran, eastern Pakistan, western Iraq, and other locations (Aghanabati 2013). The Lut Desert of Iran, for example, is extensive, and is one of the hottest places on Earth. In addition, the Hindu Kush and Karakoram Ranges in northern Afghanistan and Pakistan have peaks with elevations as great as 6000–8611 m. This area contains 13 out of the 30 highest peaks in the world (5 higher than 8000 m and the rest over 7000 m) (Kazemi and Qasim 1997). For example, the K2 peak of the Hindu Kush Range is 8611 m high and is frozen for 10 months of the year and its glaciers are world famous. The volume of ice and snow in the Karakorum Range is the largest in the world outside the polar regions. The glaciers of northern Pakistan can range from small glacial cirques to large alpine glacial valleys sometimes more than 600 m thick (Ibrahim 2004).

S.A. Aghanabati (✉)
Geological Survey of Iran, Azadi Sq., Ave. Meraje, Tehran, Iran
e-mail: agha1036@yahoo.com



A. Tsuneki et al. (eds.), *Ancient West Asian Civilization*,
DOI 10.1007/978-981-10-0554-1_3

The occupants of WA countries represent a mix of various ethnic groups, some the descendants of very ancient civilizations, including the Irano-Aryans, Turks, Arabs, Baluchis, and Kurds. As an example, the capital city of Azerbaijan, Baku, has been inhabited by various civilizations since Paleolithic times. In historic times, the twelfth century citadel of Baku was constructed over the remains of an older stronghold that was built in the sixth and seventh century B.C. The WA is the cradle of many of religions, with Islam currently the most prominent; others include Zoroastrianism, Hinduism, and Buddhism.

## 3.2 Geography

The Republic of Azerbaijan is in the Caucasus region at the crossroads of Eastern Europe and Western Asia. It has an area of approximately 86,600 km$^2$, and is bordered by Iran, Turkey, Armenia, and Georgia. The population has a mix of ethnic backgrounds, including Turkish, Armenian, Talysh, and Lezgi. Islam is the predominant religion (73 % of residents). The main geomorphologic units of Azerbaijan are the low-lying lands of the Caspian Sea shore, the Kura-Aras peneplain area, and the Great Caucasus in the north and the Lesser Caucasus in the south. About 10 % of Azerbaijan has an elevation of at least 4900 m. The main morphotectonic units of Azerbaijan are the Great Caucasus fold system, the Kura intermountain depression, the Lesser Caucasus folded mountain system, the Talesh zone, and the South Caspian depression (United Nations 2000).

Afghanistan, with a total area of 699,000 km$^2$, is a mountainous country with a mean elevation of about 1200 m. In the northeastern part of the county, the Tangeh-Vakhan heights and Hindu Kush Mountains form a strategic barrier between northern and southern Asia, with a peak height, length, and width of 7315 m, 600 km, and 100 km, respectively. Afghanistan is divided into three physiographic units, namely the northern plains, the central high terrane and the southern plateau. The high mountains are often covered with snow, but the country also has hot and dry lowlands. The Afghan people are descendants of Uzbeks and Turkmen, as well as tribes such as the Pashtu, various Persian tribes, and the Baloch (United Nations 1995).

Pakistan is part of the Indian subcontinent and is about 800,000 km$^2$ in area. The landscape is varied. In the north, there are the high peaks of the Hindu Kush and Karakoram Ranges, whereas the Indus Plain is southeast of the mountainous region, and the Thar Desert occupies much of the eastern part of Pakistan (Kazemi and Qasim 1997).

The Caspian Sea, which is a remaining part of the ancient eastern ParaTethys Ocean, is the largest lake in the world (1200 km long and 120–550 km wide). It was separated from the Black Sea and the Mediterranean during Oligocene to Miocene. The Caspian Sea bed is divided into three morphologic regions. The deepest part of the northern region is 16 m deep with a mean depth of 12 m. The central part is a great trough, which has a maximum depth of 800 m and a mean depth of 400–500 m. The southern part is the deepest region, with maximum depths of more than

1000 m. Its salinity is about 12 g/L, or about one-third of that of the open seas and oceans. About 90 million tons of sediments are deposited annually in the Caspian Sea. Sediments under the Caspian Sea contain considerable hydrocarbon reserves, which were first discovered as oil and gas springs in the fifth century (Zenkevich and Zevina 1969).

Turkmenistan has a total area of 488,000 $km^2$ and borders Kazakhstan, Uzbekistan, Iran, Afghanistan, and the Caspian Sea. The Karakoram Desert covers most of the country. The KopetDagh and Yalkhan Mountain Ranges are in the south and west of the country and have peaks with maximum heights of 2912 m and 1880 m, respectively.

Iraq is located in the northeast corner of the Arabian tectonic plate. Most of Iraq is lowland and arc-like terrane. However, a small part of northeast Iraq is a continuation of the Zagros Heights and has mountainous morphology. The Euphrates, and Tigris Rivers transfer sediments produced by weathering of the rocks from the surrounding highlands and have played an important role in formation of the Mesopotamian Plain. From the northwest, the plain dips toward the Persian Gulf by about 10 cm per kilometer. Increasing river discharge, spring storms, and winds that push the waters of the Persian Gulf landward hinder discharge of the Shatt-Al-Arab River into the Persian Gulf; all of these are contributing factors in the flooding of the Mesopotamian Plain, where as much as tens of thousands of square kilometers can be covered with floodwater 2–3 m deep during the flooding season (Jassim and Goff 2006).

Oman has an area of 300,000 $km^2$ and a population of a little more than two million. The country lies along the southeast coast of the Arabian Peninsula. The sedimentary basins of Oman, which contain important hydrocarbon reserves, cover 170,000 $km^2$.

The Arabian Peninsula has an area of more than 3 million $km^2$ and is the largest peninsula in the world. It is bordered by the Red Sea, Arabian Sea, Gulf of Aden, Persian Gulf, and Mediterranean Sea. Saudi Arabia, with a total area of 240,000 $km^2$, is the largest country on the peninsula as well as in WA.

Turkey, which is at the junction of the Asian and European continents, is a peninsula bordered by the Black Sea (north), Mediterranean Sea (south), and Aegean Sea (west). The country extends about 1600 km from east to west and 550 km from north to south. It is a mountainous area with an elevated plateau. The Ararat volcano, with an elevation of 5167 m, in the north of Turkey is the highest peak. The Euphrates (2696 km) and Tigris (1851 km) are the most important rivers.

## 3.3 General Geology

### *3.3.1 Type and Thickness of the Crust*

The Earth has a three-layer structure comprising crust, mantle, and core. The crust is the thin outer layer that is the prime subject of geological investigation. Its physico-chemical characteristics are not homogenous and can be divided into two main types, continental and oceanic, which are fundamentally different from each other in terms of their density, deformation, and thickness. As a result, their responses to various geological phenomena are not the same; for example, under compressional stress, continental crust is generally folded and faulted with an overall increase in thickness, whereas oceanic crust is usually subducted.

Continental sialic crust prevails over the vast extent of WA. However, oceanic crust lies under the Black Sea, Oman Sea, South Caspian Sea (Galperin et al. 1962), Red Sea, and Gulf of Aden (Mc Kenzie et al. 1970). Apart from these areas, allochthonous remnants of oceanic crustal rocks, which indicate the limits of ancient continents, are found along certain intra-continental fault zones. This indicates that the joining together of a number of micro-continental blocks formed the crust of WA. Based on gravimetric data, the average thickness of the continental crust is 40 km (Dehghani and Markis 1983), which increases to 60–78 km in some areas, for example, at the boundary between the Arabian and Iran plates and at the collision zone between Gondwana and Laurasia. Oceanic crust is generally less than 10 km thick.

### 3.3.2 Crustal Geodynamics and Major Tectonic Plates of WA

A large part of the WA terrane was formed during the Alpine–Himalayan Orogeny, which extended from the eastern margin of the Atlantic to the western margin of the Pacific. The area consists of several continental fragments that have joined since the late Neogene. The remnants of ophiolites (oceanic crust) and accretionary wedges that have considerable thicknesses at suture zones point towards the separation of continental fragments of WA by oceanic areas during much of the Phanerozoic. Although some such separations are believed to have been as vast as the Atlantic, most of them, including the one on the northeast part of the Arabian plate that separated the northern margin of the Gondwana supercontinent (the Iran plate), have been more or less of intra-continental origin (e.g., the Red Sea) (Nabavi 1976).

Apart from a number of small plates, such as the Central Iran microcontinent and the Pontid plate of Turkey, three super-plates have been recognized in WA (Fig. 3.1).

1. The Turan plate lies north of the Paleotethys suture, which corresponds to the margin of Laurasia.
2. The Iran plate, which comprises the terrane between the Paleo- and Neotethys that was connected to the northern margin of Gondwana until about 240 Ma (middle Triassic), was subsequently separated to join Laurasia.
3. The Arabian plate, which was first positioned on the northeastern margin of Africa, was subsequently separated from the Iran plate (in the early Mesozoic) and North Africa (in the late Neogene by the opening of the Aden and Red Sea rifts) to form the world's largest peninsula (Ziegler 2001).

A characteristic feature of the Iran plate is an abundance of igneous intrusions of various ages, which are rare in the other two plates. This phenomenon is attributed

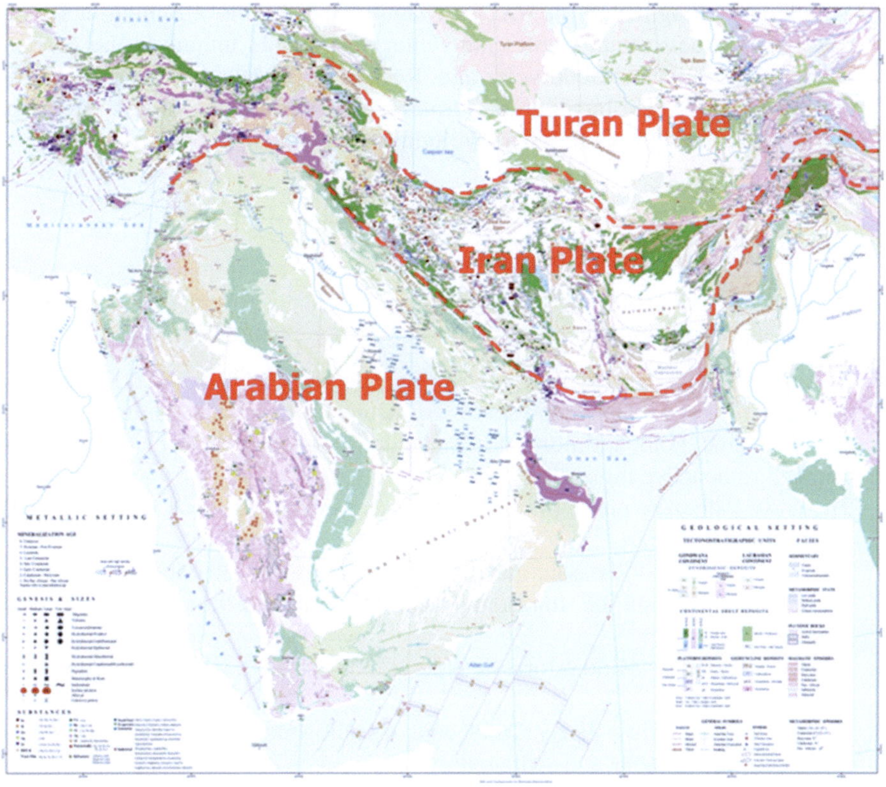


**Fig. 3.1** Three super plates of WA

to a number of factors including applied stresses from the other two plates, which have led to extensive faulting, subduction, and rifting of the Iran plate.

## 3.4 Mineral Resources of WA

The mineral resources of WA can be summarized as follows.

1. The most valuable mineral resources of WA are metals, some of which are of global importance in terms of both quantity and quality.
2. Most of the mineral resources, especially metallic resources, are closely associated with igneous activity.
3. Vast igneous activity within the Iran plate has caused a concentration of metallic resources in Iran, Turkey, Afghanistan, and parts of Pakistan.
4. Other than the Arabian Shield (near the Red Sea), which is rich in gold, silver, zinc, and copper, the western part of the Arabian plate (the Arabian Platform) is rich in mainly non-metallic minerals.

5. Mining has a long history in WA; for example, indications of mining in Turkey date back to the seventh to fifteenth century B.C. Gold mining in the Arabian Peninsula began about 5000 years ago. Results of investigations carried out by the Cultural Heritage Institute of Germany have shown that the early Iranians were pioneers in the extraction and refining of metals.

### 3.4.1 *Mineralization Phases*

The following mineralization phases have been recognized in WA.

1. The Proterozoic – early Cambrian metallogenic phase (830–517 Ma) resulted in the formation of valuable deposits of sedimentary and volcano-sedimentary iron ore, lead–zinc, gold, antimony, gold–arsenic, sedimentary phosphate, rock potash, and radioactive minerals.
2. The early Paleozoic phase (488–439 Ma) whose most precious reserve is sedimentary phosphate.
3. The late Paleozoic – Triassic metallogenic phase (299–230 Ma), during which copper–lead–zinc sulfide, tungsten, bauxite, barite, and fluorite deposits were formed.
4. The Late Triassic – Middle Jurassic phase (235–165 Ma), which was a period of coal formation.
5. The Middle Jurassic – Early Cretaceous metallogenic phase, which produced metals of both igneous and sedimentary origin.
6. The Late Cretaceous – Paleogene metallogenic phase (55.8–44.6 Ma) that gave rise to chromium, nickel, manganese, and other metals associated with oceanic crust.

### 3.4.2 *Mineral Provinces*

There are major mineral provinces in each of the Turan, Iranian, and Arabian plates as well as in the intervening rift zones.

1. Turan plate

Gigantic gas fields provide the most important economic resource in the Turan plate (Gregory 2004).

2. Iranian plate

In the Iranian plate, Precambrian–Quaternary igneous activity, especially metamorphic and metasomatic processes, have given rise to more than 95 % of WA's mineral resources (Qorbani et al. 2000).

In the Republic of Azerbaijan, there are 10 metallogenic zones and 17 ore-bearing areas with copper, molybdenum, and polymetallic sulfide deposits as well as hydrothermal gold, arsenic, cobalt, and other metals (United Nations 2000).

In Afghanistan, there are 21 metallogenic zones containing 37 mineralized areas, 1428 reserves and indications containing 10 groups of minerals. In addition to ferrous (iron, manganese, and chromium) and non-ferrous (e.g., copper, lead–zinc, aluminum, molybdenum, tungsten, tin, and bismuth) metal deposits, rare metals such as beryllium (8 areas), lithium (12 areas), tantalum, cesium, and rubidium are mined in Afghanistan. Some of these deposits are comparable, for example, to copper deposits in Zambia. Moreover, the largest copper deposit in the WA (Hajigak Mine) is in Afghanistan (United Nations 1995).

In Turkey, mining has a long history. In addition to traditional mineral resources such as lead, copper, zinc, iron, chromium, and mercury, there are more than 30 other economically valuable minerals. For example, the borax deposits of Turkey constitute more than half of the world borax reserves.

In Iran, in addition to Precambrian iron deposits formed as a result of global iron mineralization, lead–zinc and porphyry copper deposits are also found.

In northeastern Iraq, igneous activity has been responsible for the formation of many hydrothermal metallic deposits, for example, the Serguza lead–zinc deposit, the iron deposits of the Zagros suture zone, the Marapasta iron deposit, and Panjvin chromite (Jassim and Goff 2006).

3. Arabian plate

Gold extraction from volcanogenic basement in the Arabian plate dates back to 3000 B.C.; today, gold is extracted from very low-grade (<0.75 g/ton) deposits using technologically advanced methods. The mostly magmatogenic rocks of the Arabian Shield host these deposits; other gold-bearing rocks of Arabia include strata-bound massive sulfides, metasomatic hydrothermal zones, epithermal and hypothermal veins within intrusive rocks. Silver, zinc, and copper deposits are also found in the Arabian Shield.

On the Arabian Platform of the Saudi Arabia – Iraq region there are also non-metallic mineral deposits (e.g., laterite, bauxite, phosphorite, and celestite). Moreover, the concurrence of the various criteria necessary for the formation of hydrocarbons, such as source rocks, reservoir rocks, and cap rocks has made areas of the Arabian Platform (Zagros, Iraq, Arabia, Persian Gulf, Qatar, United Arab Emirates, and Oman) one of the largest hydrocarbon provinces of the world.

4. Rift-zone deposits

Rift-zone mineral deposits originate in oceanic crust and are dominated by chromite, manganese, magnesite, massive copper, and other minerals. For example, chromite deposits are found in oceanic crust in Turkey and eastern and southeastern Iran (Qorbani 2010), and there are also chromite deposits near Islambagh, Pakistan (Ibrahim 2004).

## 3.5 Energy Resources in WA

The energy resources in WA can be categorized into three main groups: coal, radioactive minerals, and hydrocarbons (natural gas and petroleum). Geothermal energy is also produced in Iran and Turkey, where the collision of tectonic plates has led to the formation of magma chambers that heat the shallow crust and form geothermal systems there. In western Turkey, a geothermal system at a depth of about 2500 m reaches a temperature of 232 °C; the geothermal heat is used to produce electricity, for commercial and domestic heating, for geo-tourism, and for medical treatments.

1. Coal

With the exception of the coal deposits of the Turan plate (e.g., Agh-Darband (Aghanabati 1999) in northeastern Iran and others in northern Afghanistan (United Nations 1995), all of the major coal deposits of WA are in the Iranian plate. Arabian plate coal deposits are limited to the Indus Plain of Pakistan (Ibrahim 2004). In Iran and Afghanistan, which cover the largest part of the Iranian plate, there are Late Triassic – Middle Jurassic economic coal deposits in foreland basins formed post early Cimmerian Orogeny. In Turkey, there are lignite and anthracite coal deposits; there is more than 8 million tons of lignite in deposits of Oligocene, Miocene, and Pliocene age, whereas the anthracite deposits are estimated at 1.1 billion tons. Despite these resources, Turkey imports 45 million tons of coal annually (e.g., Aghanagati 2013).

Pakistan has vast resources of low-quality lignite and bituminous coal of Tertiary age in the Indus Basin; the most important of these are in the Thar Coalfields, where estimated reserves are 184 billion tons, which stands amongst the largest known lignite deposits in the world. Overall, the lignite reserves of Pakistan total 630 billion tons (Ibrahim 2004).

2. Radioactive elements

Afghanistan, Pakistan, and Turkey are known to have deposits of radioactive minerals, but no data are available for other WA countries. In the Sulaiman Range of Pakistan, there is a band of uranium-bearing rocks within which there are seven intense anomalies with an average assay of 0.138 % $U_3O_8$ (Ibrahim 2004). The radioactive minerals are calcium and uranium (decayed) hydrate (tyuyamunite), which forms a greenish yellow crust on sand grains within a sand layer between two shale layers. Radioactive and rare earth element deposits are found in the Kaneshin region in Helmand Province of Afghanistan, but they are confined to early Quaternary volcanic carbonatites.

3. Hydrocarbons

Oil and natural gas accumulations were first recognized as surface leakages and were exploited as natural flows. For example, the "Immortal fire flames" of Azerbaijan were considered to be a consequence of the ignition by humans of oil-bearing shales near Baku. The first oil exploration well was drilled in 1745 in France and the first oil production well was drilled in 1859 in Pennsylvania (USA). These events were the starting point of modern subsurface oil exploration. In WA, exploration activities started in 1872 when oil was discovered near Masjed-Soleyman in Iran. The use of modern technology during the past few decades has considerably increased the efficiency and success rate of oil exploration. Of the 32,000 known oilfields in the world, more than 1500 are in WA.

4. Oilfields in WA

On the southern margins of the Turan plate (Azerbaijan, Caspian Sea, Turkmenistan, and northern Afghanistan) and in extensive parts of the Zagros Arabian Plate (Iraq, Kuwait, Arabia, Qatar, Dubai, Abu Dhabi, and Oman), there are gigantic reserves of hydrocarbons; more than half of the world's reserves have been identified in this area. The processes of oil formation are fundamentally different in the Turan and Arabian plates, especially from the point of view of age. The hydrocarbon systems of the Arabian plate are of Paleozoic, Mesozoic, and Cenozoic age (Vail et al. 1977) whereas those of the Turan Plate are of Pliocene age (Aghanabati 2013).

Oil exploration began in the Republic of Azerbaijan in 1871 and has continued ever since, resulting in the drilling of more than 21,000 boreholes in various exploration areas, including the Absheron Peninsula, Caspian continental shelf, and Kura Embayment. The estimated recoverable petroleum and gas reserves of Azerbaijan amount to 9.6 billion tons; the gas reserve of the largest accumulation is 1 trillion cubic meters and the total gas reserves are estimated to be 2–7 trillion cubic meters.

The ancient Iranians considered fire sacred; they called the burning natural gas springs "Eternal Flames," and constructed fire temples around them. As noted previously, the earliest exploration for petroleum commenced in Iran 1872. Currently, there is well-documented evidence of hydrocarbon reserves in the southern Caspian Sea, and in an area on the southern flanks of the Zagros Mountains, which constitutes one of the world's largest petroleum basins. Iran has reserves of 130.7 billion barrels of oil representing 12 % of global reserves. The gas reserves of Iran are estimated at 26.96 trillion cubic meters, which constitutes 15.2 % of global reserves.

The hydrocarbon reserves of Afghanistan are mostly in gas fields that are an extension of those in Turkmenistan (see below). The estimated reserves in Afghanistan are about 2 billion cubic meters. The largest hydrocarbon-bearing basin in Pakistan is the Indus Basin, which contains 72 gas fields and 24 mostly small oilfields. The extractable gas reserves of Pakistan stand at more than 1.2 trillion cubic meters (Ibrahim 2004). The 4000-km$^2$ Amu Darya Basin of Turkmenistan (Fig. 3.2) is the largest hydrocarbon basin in the country; it extends southwards into Iran and Afghanistan. Amu Darya is a deep basin (4500–10,000 m) with passive margin deposits, notably the Gurdek Formation (J3) and Shatlik Formation (K1), which are key components of the hydrocarbon systems there. Almost 2 billion barrels of petroleum and 6.5 trillion cubic meters of gas have been identified in 250 hydrocarbon fields in the Amu Darya Basin, half of which are economically viable (Gregory 2004).

**Fig. 3.2** Political boundary and gas field of Amu-Darya Basin (Dyman et al. 2001)

International oil companies have carried out extensive exploration in Saudi Arabia since 1933, resulting in the discovery of gigantic oilfields and making Saudi Arabia the largest producer of petroleum worldwide. Arabian hydrocarbon reserves are found in three regions (Gotnia, the Persian Gulf, and the Southern Persian Gulf) and constitute 26 % of global reserves. Qatar, Abu Dhabi, and Dubai have oilfields similar to those of Saudi Arabia, but those of Oman (64 onshore and several offshore fields) are different. Unlike the Arabian oilfields, Oman petroleum accumulations are of Proterozoic age and are associated with kerogen-bearing rock salt (similar to oilfields in Siberia and Australia) (Visser 1991).

Since the time of the Sumerians (ca. 4000 B.C.), natural tar has been used for waterproofing of sea vessels and making mortar in the area that is now Iraq (Jassim and Goff 2006). Oil was discovered in a well drilled in Iraq in 1919. According to OPEC's Annual Report, during 1964–2004, 115 billion barrels of petroleum and 3.2 trillion cubic meters of gas were discovered in four fields (Kirkuk, Rumaila, Zubair, and Bihassan) in Iraq. About 50–60 million tons of petroleum and 14–15 billion cubic meters of gas are produced in the Caspian Sea region annually (Fig. 3.3).

## 3.6 Conclusion

Geological Science, as one of the most fundamental branches of the "Sciences", provides not only high productivity for habitual development, it also has a comprehensive relationship with social and cultural sciences. In other words well cultured are those who could/can realize the differences in the natural earth's capacities.

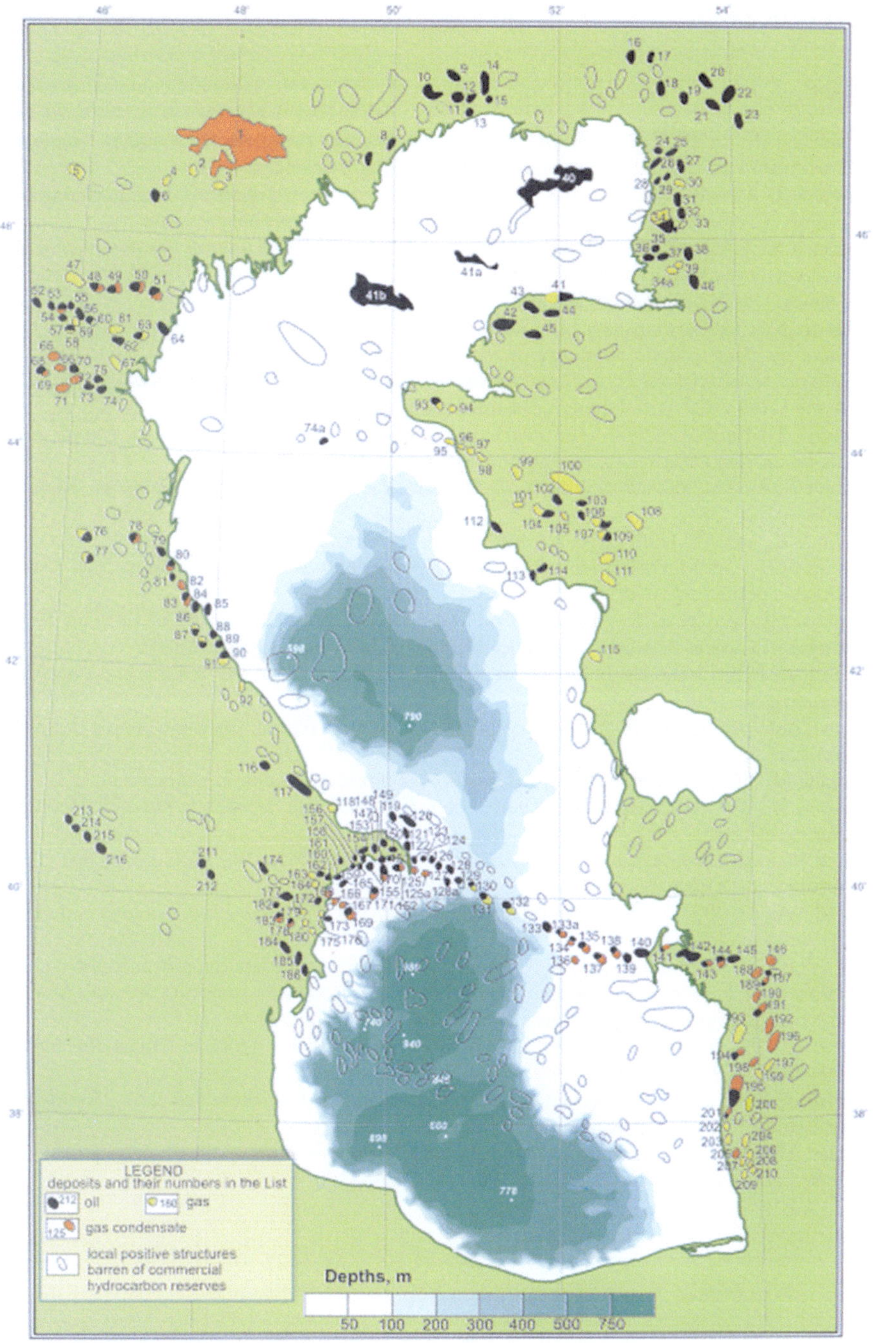


**Fig. 3.3** Oil and gas fields of the Caspian Sea (Khan and Bogdanov 2006)

## References

Aghanabati, S.A. 1999. *Jurassic in Iran (in Persian)*, Vol 2, 740. Tehran: Geological Survey of Tehran, Iran.

Aghanabati, S.A. 2013. *Geology of Iran and neighboring countries (in Persian). Geological Survey of Iran*, 710. Tehran: Research Institute of Earth Sciences.

Dehghani, G.A., and J. Markis. 1983. The gravity field and crustal structure of Iran. In *Geodynamic project (Geotraverse) in Iran. Geological Survey of Iran Report* No.51.51-68

Dyman, T.S., V.A. Litinsky, and G.F. Ulmishek. 2001. *Geology and natural gas potential of deep sedimentary basins in the Former Soviet Union*, U.S. Geological Survey Digital Data Series 67. Denver: USGS Information Services.

Galperin, E.I., I.P. Kosminskaya, and R.M. Krakshina. 1962. *Main characteristics of deep registered by deep seismic sounding in the central part of the Caspian sea from DSS data, Glubinnoeseismicheskoezondirovaniezemnoikory v SSSR (Deep Seismic Sounding of the Crust in the USSR)*, 227–250. Leningrad: Gostoptekhizdat.

Gregory, F.Ulmishek. 2004. *Petroleum geology and resource of the Amu-Darya Basin, Turkeminstan, Uzbakestan, Afghanestan and Iran*, 32

Ibrahim Shah, M. 2004. Stratigraphy of Pakistan. Geological Survey of Pakistan (Karachi.). *Memoir* 12: 138.

Jassim, S.Z., and J.C. Goff. 2006. *Geology of Iraq*, 241. Prague: Dolin.

Kazemi, A.H., and M. Qasim Jan (eds.). 1997. *Geology and tectonics of Pakistan*, 554. Karachi: Graphic publisher.

Khan, V.Y.E., and N.A. Bogdanov. 2006. *International tectonic map of the Caspian Region*. Moscow: Russian Academy of Science Geological Institute, Moscow: Russia.

Mc Kenzie, D.P., D. Davies, P. Molnar, et al. 1970. Plate tectonic of the Red Sea and East Africa. *Nature* 226: 243–248.

Nabavi, M.H. 1976. *An introduction to geology of Iran*, 109. Tehran: Geologic Survey of Iran. (In Persian)

Qorbani, M. 2010. *Economic geology of Iran*, 449. Tehran: Arian Zamin Publication.

Qorbani, M., Tajbakhsh, P., and N. Khoie. 2000. *Lead and zinc deposits in Iran*, 75. Geological Survey of Iran.

United Nations. 1995. *Atlas of mineral resources of the ESCAP region*, 85. Geology and mineral resources of Afghanistan, 11

United Nations. 2000. *Atlas of mineral resources of the ESCAP region*, 216. Geology and mineral resources of Azarbaijan, 15.

Vail, P.R., R.M. Mitchum Jr., and S. III Thompson. 1977. Seismic stratigraphy and global change of sea level. In *Seismic stratigraphy application to hydrocarbon exploration*, Memoir American association of petroleum geologists, vol. 26, ed. C.E. Payton, 83–97.

Visser, W. 1991. Burial and thermal history of Protrozoic source rocks in Oman. *Precambrian Research* 54: 15–36.

Zenkevich, L.A., and G.B. Zevina, eds. 1969. *Fauna and flora, Caspian Sea*, 229–255. Moscow: Moscow University. (in Russian)

Ziegler, M. 2001. Late permian to holocene paleofacies evolution of the Arabian plate and its hydrocarbon occurrence. *GeoArabia* 6(3): 445–504.

# Chapter 4
# Paleoclimatic Changes and Human Cultural Evolution in West Asia

**Ryo Anma and Teruyuki Maruoka**

Recent studies of paleoenvironmental changes in and around the Arabian Peninsula have suggested that extremely wet periods occurred during the late Pleistocene and early Holocene; these periods started around the termination of the glacial periods and continued during the initial rapid warming that followed. During these periods, floodwaters transported enormous amounts of sediments from mountains to plains. Once the thermal maximum was reached and temperature stabilized at a high level, the water environment also stabilized and the climate became dryer. Although gradual changes in atmospheric temperature may have had only a minor direct influence on human activities, drastic changes in the water environment in response to changes in the average atmospheric and surface water temperatures of just a few degrees had a stronger influence on human culture, possibly leading to the development of agriculture and civilizations. Domestication of animals started at the beginning of a wet period in the early Holocene, whereas the first cities were constructed during its latest stage. The need to control flooding may have led humans to build cities and, thus, to the development of early civilizations.

## 4.1 Introduction

The modern human species (*Homo sapiens*) evolved in East Africa from their ancestors under climatic conditions that changed frequently from warm to cold and back to warm, but with a trend toward dryer and cooler conditions during the last 4 million years (My)(deMenocal 2004). The period of these warm–cold cycles changed from 19 to 23 thousand years (ky) to 41 ky around 3 million years ago (Ma), and

R. Anma (✉) • T. Maruoka
Faculty of Life and Environmental Sciences, University of Tsukuba,
1-1-1, Tennodai, Tsukuba 305-8572, Ibaraki, Japan
e-mail: amma.ryo.ge@u.tsukuba.ac.jp



A. Tsuneki et al. (eds.), *Ancient West Asian Civilization*,
DOI 10.1007/978-981-10-0554-1_4

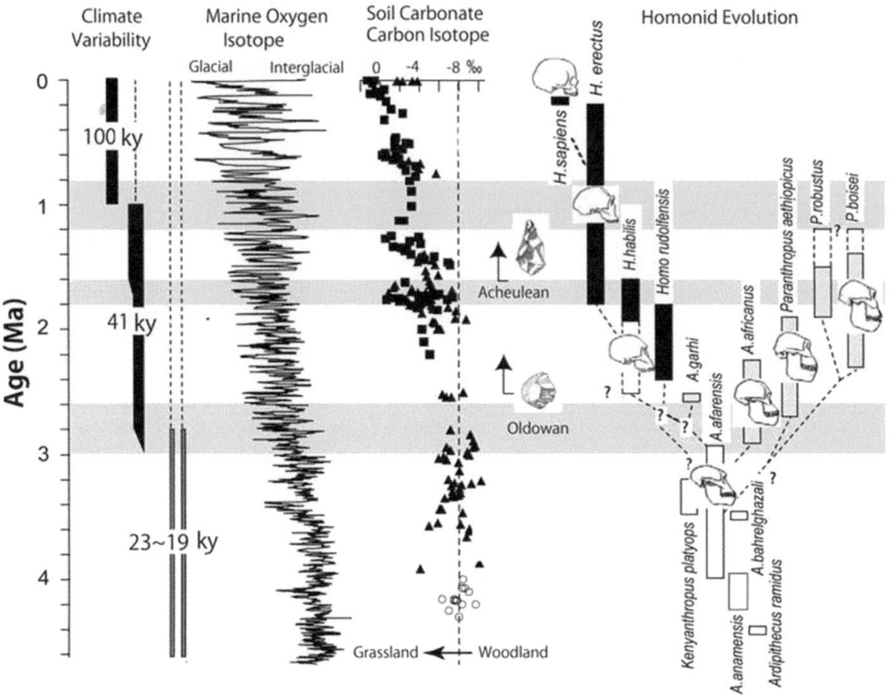


**Fig. 4.1** Important Pliocene–Pleistocene paleoclimatic and hominid evolution events (modified from deMenocal 2004 with permission by the author). African climate variability data are from deMenocal (1995); marine oxygen isotope data are from Shackleton et al. (1990) and Mix et al. (1995); East African soil carbon isotope data are from Cerling and Hay (1988), Cerling (1992), and Wynn (2000); and hominid evolution data are from Lieberman (2001)

from 41 ky to 100 ky at ~1 Ma; the amplitude of the climatic changes increased with time and became larger after 1 Ma. Since 1 Ma, the onset of cooling has occurred gradually whereas warming has generally taken place rapidly. Anthropoids and modern humans evolved under these repeated gradual cooling and rapid warming cycles in almost the same areas in East Africa (Fig. 4.1).

Migration of *H. sapiens* "out of Africa" was once thought to be only possible via the Sinai Peninsula and the Levant corridor, during warm periods and perhaps that was the case. More recently, however, excavations in the Arabian Peninsula have found evidence that *H. sapiens* reached the western peninsula via its southern coast as early as 125 ka (Armitage et al. 2011). Paleoenvironmental studies in the Red Sea have revealed that, because of the global sea-level fall during the last glacial maximum (LGM), a land bridge existed across the Bab-el-Mandeb, the strait connecting the Gulf of Aden with the Red Sea (Fenton et al. 2000), which allowed humans to migrate across the strait. At the same time, the Persian Gulf, which lies to the north of the peninsula, was a shallow water body, or it may have completely dried up. Similar land bridges may have existed during previous cold periods, providing anthropoids and modern humans a pathway by which to migrate overland from East

Africa and across the Arabian Peninsula to Persia, and thence to the whole Eurasian continent and throughout the world. The first agricultural practices and human civilizations were developed along this migration pathway in western Asia.

In this chapter, we focus on climate changes in western Asia and consider how they influenced human cultural development. On the basis of a review of paleoenvironmental studies conducted in and around the Arabian Peninsula, we suggest that changes in precipitation may have had extensive influence on the development of agriculture and civilizations. Frequent floods during wet periods, such as those recorded in ancient writings such as the Epic of Gilgamesh, and the need to construct flood control works may have led humans to organize large labor forces, leading in turn to the foundation of cities and the development of civilizations.

## 4.2 Paleoclimatic Changes and Wet Arabia

Our image of the landscape of western Asia may correspond to the barren, dry, yellowish ergs of Arabia, or the rocky deserts of eastern Iran. As viewed from space, a dry, yellowish belt between about 20 and 30°N latitude extends from the Sahara Desert in North Africa to eastern Iran (Fig. 4.2) and covers the southern part of western Asia. However, the present dry, barren yellow belt is just a snapshot of the face of an Earth that is changing through time with changing climatic conditions. A

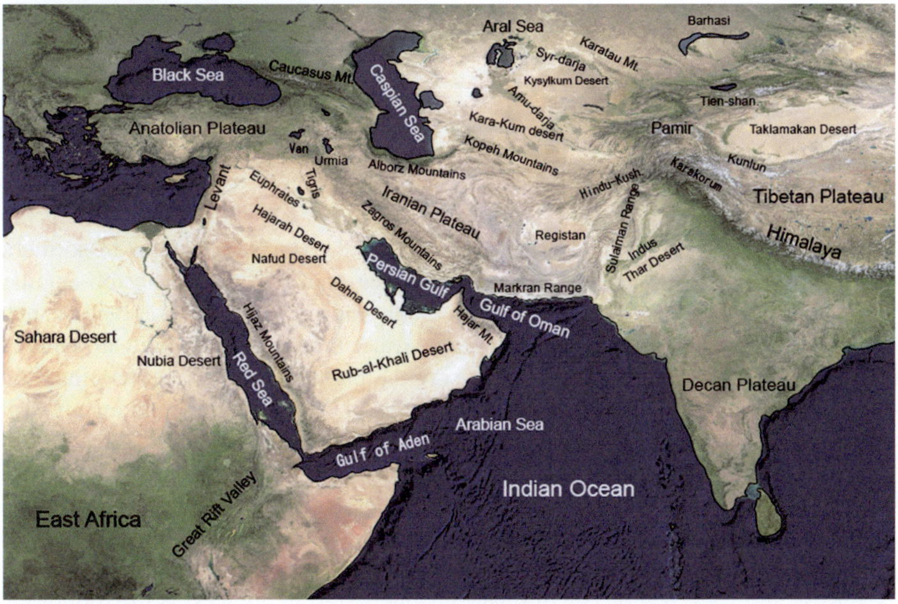


**Fig. 4.2** Topography of West Asia (based on a Google Earth image). The area from 50°N latitude to the equator and between 20°E and 90°E longitude is shown

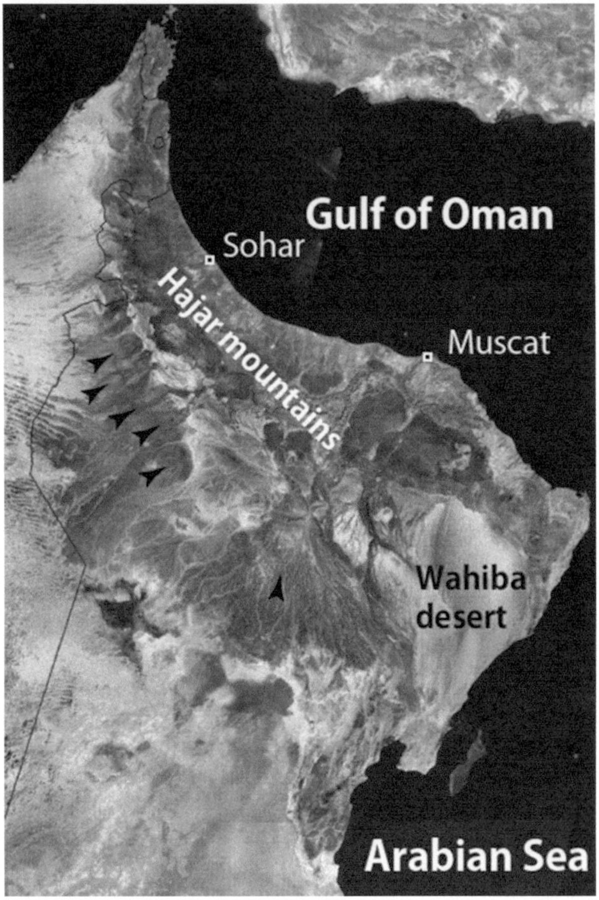


**Fig. 4.3** Satellite image of the Hajar Mountains, Oman. *Arrows* indicate fan deposits (Modified from a color image provided by the Geological Society of Oman (2006))

satellite image of the eastern Arabian Peninsula (Fig. 4.3) shows a number of fan deltas along the western and southern margins of the Hajar Mountains in northern Oman. These thick alluvial deposits consist primarily of sub-rounded to sub-angular boulders, commonly up to a half meter in diameter (Fig. 4.4). The presence of such deposits implies a high-energy, high-volume water flow, and hence high amounts of precipitation in the Hajar Mountains, where the present mean annual precipitation is less than 300 mm (Glennie and Singhvi 2002). The Hajar Mountains are still within the monsoon rainfall belt (Fig. 4.5), so these sediments could have been deposited even in today's desert climate by surface runoff flowing at a high rate. However, beneath the Rub-al-Khali, the driest desert of Arabia with a mean annual precipitation of less than 50 mm, drainage channels, alluvial conglomerates, and even lake sediments containing shells and microfossils are widely distributed (Fig. 4.5; Rose and Usik 2009). The drainage channels start in the west at the foot of the Hijaz Mountains and extend across the entire peninsula from west to east, entering the Gulf of Oman via the basin of the modern Persian Gulf. Thus, this region of the Arabian Peninsula was once well watered.

**Fig. 4.4** Thick alluvial deposits observed along a gorge in the foothills of the Hajar Mountains

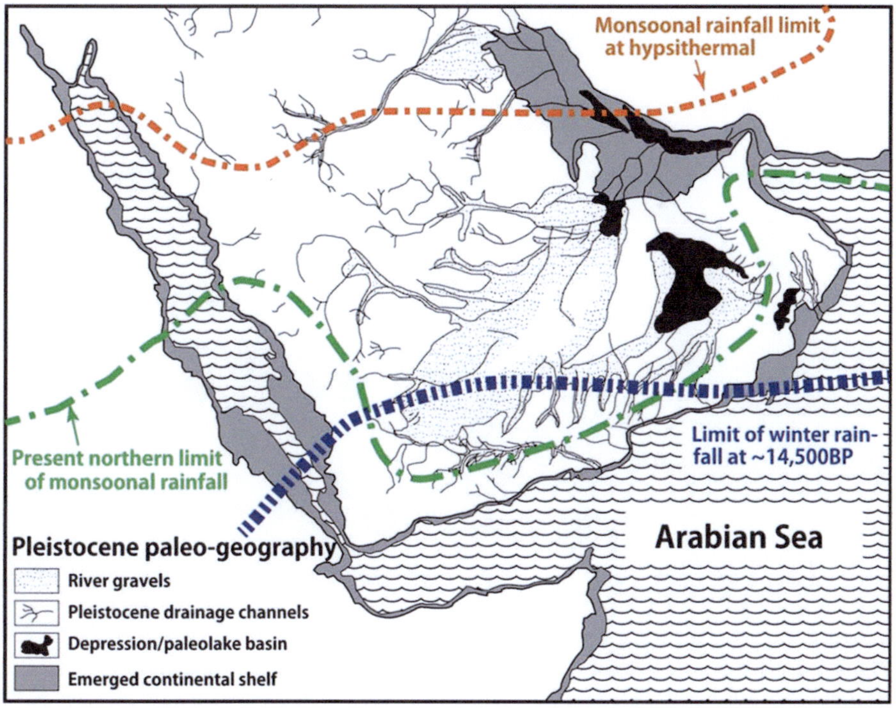


**Fig. 4.5** Pleistocene drainage systems and paleogeography during the last glacial maximum (modified from Rose and Usik 2009 with permission from Springer). The present northern limit of monsoonal rainfall (summer rainfall) is from Boivin et al. (2009); the limits of monsoonal rainfall during the hypsithermal and of winter rainfall at ~14,500 y BP are from Butzer (1995)

## 4.3 Late Pleistocene – Holocene Environmental Changes in Arabia

To examine the environmental changes that have occurred since the late Pleistocene in the Arabian Peninsula and surrounding region, we compiled the results of paleoenvironmental studies from sites distributed in and around the peninsula (Fig. 4.6). In this paper our focus is on paleoenvironmental changes since 140 ka, or marine isotope stage (MIS) 6 (Fig. 4.7), which includes the two prominent interglacial periods, the last interglacial (MIS 5e) and from 10 ka to the present in the Holocene (MIS 1), and two major glacial periods, the LGM, at ~20 ka in MIS 2, and MIS 6.

The last interglacial was as warm as or even warmer than the Holocene, and the global sea level (Fig. 4.7a) was as high as it is at present. As the climate cooled toward the LGM, vast amounts of seawater became locked up in the polar ice caps and continental glaciers, and global sea level dropped by ~120 m. During MIS 6,

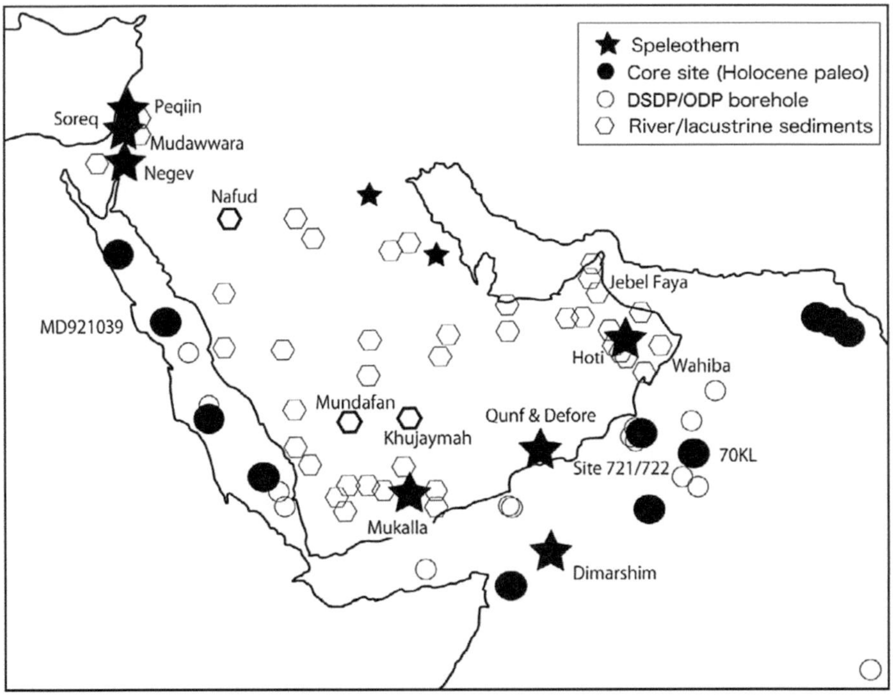


**Fig. 4.6** Distribution of paleoenvironmental study sites on the Arabian Peninsula. The labeled sites are those included in Fig. 4.7. Data for speleothems and Holocene core sites are from Fleitmann et al. (2007) and Rosenberg et al. (2013); lacustrine and river sediment data are from Parker (2008) and Butzer (1995); Ocean Drilling Program (ODP)/Integrated Ocean Drilling Program (IODP) core site data are from the IODP home page (http://iodp.tamu.edu/scienceops/maps.html)

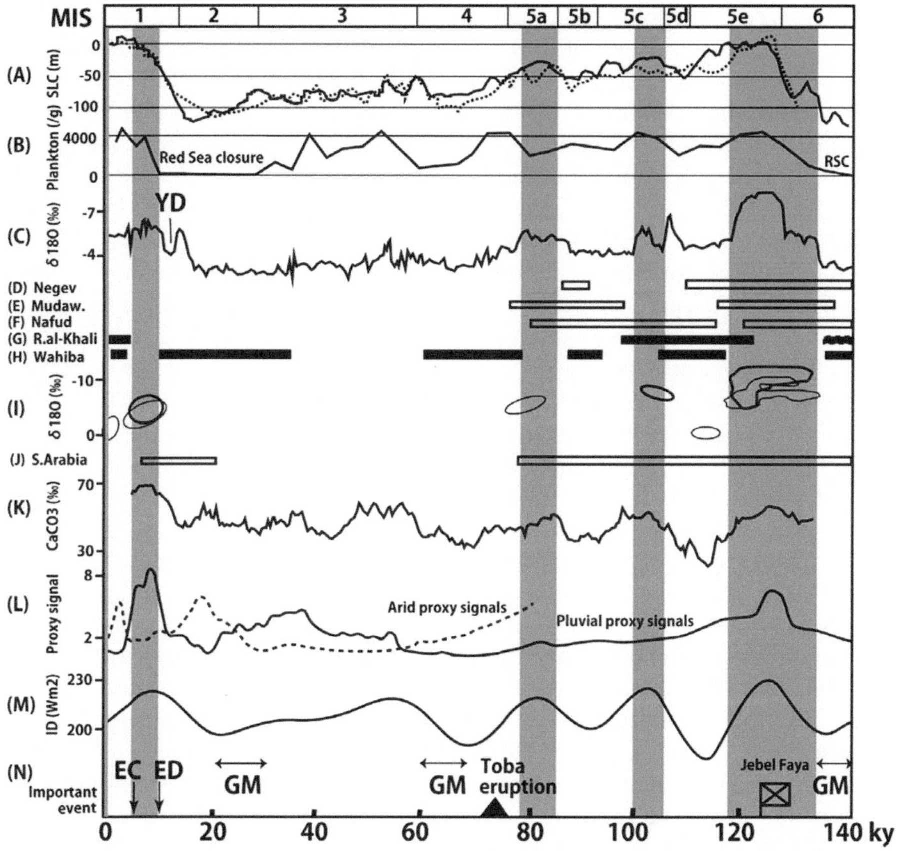


**Fig. 4.7** Paleoenvironmental changes since 140 ka. Shaded vertical bars represent wet periods. (**a**) MIS stages and sea-level changes. *Solid line*, global sea-level changes (http://www.ncdc.noaa.gov/paleo/ctl/clisci100k.html); *dotted line*, Red Sea sea-level changes (Bailey 2009). (**b**) Changes in planktonic foraminiferal content (number per gram) in a Red Sea core (Fenton et al. 2000). *RSC* Red Sea closure. (**c**) Oxygen isotopic ratios from Soreq cave speleothems, central Israel (Bar-Matthews et al. 1999, 2003). *YD* Younger Dryas. (**d–h, j**) *Solid bars* are for a dry proxy; *open bars* are for a wet proxy. (**d**) Th/U ages of speleothems in the central and southern Negev Desert (Vaks et al. 2010). (**e**) Th/U ages of coquina shells from the Mudawwara lacustrine deposit in southern Jordan (Petit-Marie et al. 2010). (**f**) OSL and TT-OSL ages on lacustrine sediments in the Nafud Desert (Rosenberg et al. 2013). (**g**) OSL ages of sand dunes in Rub-al-Khali (Preusser 2009) and (**h**) the Wahiba Desert (Preusser et al. 2002). (**i**) Th/U ages and oxygen isotopic ratios of speleothems from Hoti cave, Oman (*thick line*; Fleitmann and Matter 2009) and Mukalla cave, southern Yemen (*thin line*; Fleitmann et al. 2011). (**j**) OSL ages of lacustrine deposits from Mundafan and Khujaymah (Rosenberg et al. 2011). (**k**) $CaCO_3$ contents of core 70KL from the Arabian Sea (Leuschner and Sirocko 2000). (**l**) Humid proxy signal (*solid line*) and arid proxy signal (*dashed line*) (Parker 2008). (**m**) Indian summer monsoon index (Leuschner and Sirocko 2003). (**n**) Important events: GM, glacial maximum; *ED* early domestication of animals (Bar-Yosef 1998); *EC* birth of cities (Pollock 1999); eruption of Toba volcano (Westgate et al. 1998)

sea level was slightly lower than the LGM sea level. Local sea-level changes in the Red Sea, deduced from stable oxygen isotope ratios of planktonic foraminifera (dotted line in Fig. 4.7a Bailey 2009), followed the general trend of the global sea-level changes. In a detailed study of Red Sea marine cores, Fenton et al. (2000) identified an interval in the marine core sediments that contain no or very few planktonic foraminifera (Fig. 4.7b), called the aplanktonic zone, that covered the entire LGM. They suggested that, owing to the sea-level fall, the Bab-el-Mandeb had closed at that time, which allowed the salinity of Red Sea waters to rise to levels too high for planktonic foraminifera to survive. A similar aplanktonic zone was observed in MIS 6. Thus, during both periods, a land bridge likely connected the Great Rift Valley, the cradle of hominid evolution, with the Arabian Peninsula. The MIS 6 land bridge would have provided a pathway for *H. sapiens* to reach Jebel Faya, the eastern tip of the Arabian Peninsula, as early as 125 ka (Armitage et al. 2011). In addition, most, if not all, of the Persian Gulf must have been dry land at that time, thus providing a highway for humans migrating beyond the Arabian Peninsula to other parts of western Asia.

Terrestrial paleoenvironmental changes have been studied using various proxies. For example, actively migrating sand dunes are an indicator of a dry climate. Sand dunes can be dated by applying optically stimulated luminescence (OSL), infrared stimulated luminescence (IRSL), and thermally transferred (TT)-OSL dating methods to sand grains composed mainly of quartz. When the sand dunes are actively migrating, the OSL and IRSL signals are continuously reset. In this way, Preusser et al. (2002, 2005) and Radies et al. (2004) demonstrated that the sand dunes in the Wahiba desert (Fig. 4.3), eastern Oman, were generally active during cold periods (Fig. 4.7h), and Preusser (2009) showed that dunes in Rub-al-Khali were active during MIS 5d and the late Holocene, as they are at present (Fig. 4.7g).

Proxies indicating a wet climate include the deposition of lacustrine and alluvial river sediments. Radies et al. (2005) obtained IRSL ages of sediments and $^{14}C$ ages of shells in lacustrine deposits of the Wahiba Sand Sea and showed that this area was wet from around 11 to 8 ka. Petit-Marie et al. (2010) used alpha spectrometry to obtain the distribution of Th/U ages of *Cerastoderma* shells (coquina) from lacustrine deposits in the Mudawwara depression, southern Jordan (Fig. 4.7e). They showed that there was a paleolake in the depression from late MIS 6 to MIS 5e, and again around MIS 5b. Rosenberg et al. (2013) showed by using both conventional OSL and TT-OSL that the lacustrine sediments in the western Nafud Desert were deposited around 125 ka and around 100 ka (Fig. 4.7f). Lacustrine deposits also accumulated in Mundafan and Khujaymah, southern Arabian Peninsula, in MIS 6 and throughout MIS 5, and again from the LGM to the early Holocene (Fig. 4.7j; Rosenberg et al. 2011).

In caves, speleothems are excellent markers of wet periods. Speleothems have great advantages over other proxies, because they provide continuous data on both age, from Th/U dating, and the paleoenvironment, from the stable oxygen isotope record. Bar-Matthews et al. (1999, 2003) obtained an excellent, continuous record of oxygen isotopic ratios covering the entire time span considered from speleothem in Soreq cave, central Israel (Fig. 4.7c). The pattern of oxygen isotope variations is similar to that of the global sea-level changes (Fig. 4.7a), which suggests that the isotopic ratiosreflect paleotemperature changes.

In contrast to the speleothems in Soreq cave, which grew continuously owing to moisture from the Mediterranean Sea, speleothem growth in the southern Arabian Peninsula has not been continuous. Th/U ages of speleothems in caves in the central and southern Negev Desert (Fig. 4.7d Vaks et al. 2010) and $^{230}$Th ages and oxygen isotopic ratios of speleothems from Hoti cave in Oman (Fig. 4.7i thick line; Fleitmann and Matter 2009) and Mukalla cave in southern Yemen (Fig. 4.7i thin line; Fleitmann et al. 2011) indicate that the growth of speleothems was episodic, occurring only during wet periods. It is noteworthy that at present, oxygen isotopic ratios of rainwater in the eastern Mediterranean region range from +2‰ to −12‰ (during 2000–2002: Bar-Matthews et al. 2003) relative to Vienna Standard Mean Ocean Water (VSMOW), with annual averages ranging from −5.65‰ to −6.9‰ depending on annual precipitation. In contrast, monsoon rains in southern Oman have isotopic ratios ranging from −1‰ to +2‰ relative to VSMOW (Fleitmann and Matter 2009). Recent oxygen isotopic ratios of speleothems reported from Soreq cave are similar to those of rainfall in the eastern Mediterranean, whereas those of Hoti and Mukalla caves correspond to the ratios of Arabian Sea monsoon rains. During the hypsithermal, the warmest period in the Holocene at ~7 ka and also referred to the Holocene Climate Optimum, the limit of monsoonal rainfall was further northward than the present northern limit (Fig. 4.5). In contrast, during ~14.5 ka, the southern limit of winter rainfall occurred in the southernmost Arabian Peninsula. The changes in the oxygen isotope in the caves in Oman reflect influences of changing water masses with different oxygen isotopic ratios that supplied moisture to the southernmost Arabian Peninsula, whereas those of Soreq cave in north may have been controlled solely by the moisture from the Mediterranean Sea throughout the considered period.

Evidence for a changing terrestrial environment is also recorded in marine cores. deMenocal (2004) reported that the terrigenous input to marine sediments at ODP Site 721/722 offshore East Africa generally increased during cold periods and decreased during warm periods. In addition, changes in the sea surface temperature (and monsoon strength) also influences vertical seawater circulation and primary productivity in the Indian Ocean. Increases in primary productivity lead to increases in carnivorous microfossils such as foraminifera, which appear as increases in the $CaCO_3$ contents (Fig. 4.7k), for example, in the 70KL core from the Arabian Sea (Leuschner and Sirocko 2000). There seems to be no significant disagreement nor delay time in between the environmental proxy signals obtained from northern (Fig. 4.7c–f) and southern Arabian Peninsula (Fig. 4.7g–k).

We identified several distinctive wet periods (shaded areas in Fig. 4.7) on the basis of these paleoenvironmental records and proxies for the Arabian Peninsula. This interpretation is in coincides with previously reported probability density function curves for humid or arid climate proxy signals constructed based on 427 absolute dates of fluvial and lacustrine deposits, speleothem and marine core samples from across the Arabian peninsula (Fig. 4.7l; Parker 2008), and also with the Indian summer monsoon index calculated from the insolation difference between 30°N and 30°S (Fig. 4.7m; Leuschner and Sirocko 2003). High-precipitation periods in western Asia seem to start at the end of glacial periods and to continue as the climate warms but they end when the temperature maximum (hypsithermal in the Holocene) is reached. The remainder of the warm period, following the temperature maximum, was dry, like the climate of present-day western Asia. An important aspect is that the simple correspondences of warm period = wet period and cold period = dry period are not necessarily true. However, the most humid periods apparently occur when $H_2O$ being stored in the polar ice caps and continental glaciers is returned to the ocean and atmosphere.

An interesting event that may have influenced the global climate took place at ~75 ka, just before the MIS 4 glacial maximum (Fig. 4.7n); the eruption of Toba volcano, Indonesia, distributed volcanic ash as far as the Indian Ocean. Such a large eruption likely caused temporary global cooling by reducing the amount of solar irradiation that reached the Earth's surface. However, the actual consequences of this eruption are unknown.

## 4.4 Climatic Changes and Their Impact

In this section, we focus on the question that is of most interest to archaeologists, that is, the influence of climate change on human cultural development. The compiled paleoenvironmental data from the Arabian Peninsula presented in Sect. 4.3 demonstrates that, even though most of the Arabian Peninsula is now arid and barren, it has experienced several distinctively wet periods since 140 ka. Sea-level lowering during the MIS 6 and the LGM enabled humans to migrate to western Asia via land bridges. People who migrated to the Arabian Peninsula during MIS 6 reached Jebel Faya by 125 ka, so they might have witnessed the drastic environmental change from a wet period to a dry period. People who did not migrate from Jebel Faya across the dry Persian Gulf basin during MIS 6 had another chance to do so during the LGM. People that migrated to the Eurasian continent during the LGM must have experienced the Younger Dryas, a cold, dry period, followed by an extremely wet period as the hypsithermal approached, and eventually the dry climate of the late Holocene.

Gradual changes in atmospheric temperature likely had only a minor direct influence on human activities, because humans can adapt much faster than such gradual changes occur. However, drastic changes in the water environment in response to average atmospheric and surface water temperature changes of a few degrees can pose a direct threat to life and thus might have seriously affected agricultural activities and the development of civilizations. Early domestication of sheep and goats started during the early Pre-Pottery Neolithic B (PPBN) in Levant (Bar-Yosef 1998) and the first crop cultivation may have started ~11 ka (Tanno and Maeda, this volume), concurrently with the initiation of the early Holocene wet period (Fig. 4.7n), and the birth of cities (Pollock 1999; also see the paper by Ur, this book) coincided with the latest stage of this wet period (Fig. 4.7n). These correspondences suggest that the changes in precipitation might have had a large, direct influence on the development of agriculture and the growth of civilizations. We hypothesize that frequent heavy flooding made a large labor force necessary to construct flood control works, and that necessity may have led people to gather together in early cities, which led in turn to the development of early civilizations.

The post-hypsithermal warm period must also have been dry, like present-day western Asia. This hot, dry period may have modified the styles of the early civilizations that developed during the rainy period. In western Asia, many houses and other buildings are constructed from adobe (dried mud) and are sensitive to environmental changes such as dry-wet cycles. Thus, the influence of such climate changes must be enhanced in western Asia compared with other parts of the world. Repeated environmental changes that forced humans to adapt to the changing environment may have enhanced human cultural development and the growth of civilizations.

## 4.5 Summary

Perhaps climate change did have an influence on human cultural development. Drastic changes in the water environment, which can occur in response to changes in average atmospheric and surface water temperatures of just a few degrees, directly threaten human lives. Therefore, such changes would have greatly influenced the development of agriculture and the growth of civilizations.

We emphasize here that the simple correspondences of warm period = wet period and cold period = dry period scheme may not be true. High-precipitation periods in western Asia began at the end of glacial periods and continued during the following warming period and only ended when the temperature maximum (hypsithermal during the Holocene) was reached. Following the temperature maximum, conditions were dry, as they are in present-day western Asia. The hot, dry climatic conditions may have led to reformation of the civilizations that had developed during the rainy period.

## References

Armitage, S.J., S.A. Jasim, A.E. Marks, A.G. Parker, V.I. Usik, and H.-P. Uerpmann. 2011. The southern route "Out of Africa": Evidence for an early expansion of modern humans into Arabia. *Science* 331: 453–456.

Bailey, G. 2009. The Red Sea, coastal landscapes, and Hominid dispersals. In *The evolution of human populations in Arabia*, ed. M.D. Petraglia and J.I. Rose, 15–37. Dordrecht: Springer.

Bar-Matthews, M., A. Ayalon, A. Kaufman, and G.J. Wasserburg. 1999. The Eastern Mediterranean paleoclimate as a reflection of regional events: Soreq cave, Israel. *Earth and Planetary Science Letters* 166: 85–95.

Bar-Matthews, M., A. Ayalon, M. Gilmour, A. Matthews, and C.J. Hawkesworth. 2003. Sea– land oxygen isotopic relationships from planktonic foraminifera and speleothems in the Eastern

Mediterranean region and their implication for paleorainfall during interglacial intervals. *Geochimica et Cosmochimica Acta* 67: 3181–3199.

Bar-Yosef, O. 1998. The Natufian culture in the Levant, Threshold to the origins of agriculture. *Evolutionary Anthropology* 6: 159–177.

Boivin, N., R. Blench, and D.Q. Fuller. 2009. Archaeological, linguistic and historical sources on ancient seafaring: A multidisciplinary approach to the study of early maritime contact and exchange in the Arabian peninsula. In *The evolution of human populations in Arabia*, ed. M.D. Petraglia and J.I. Rose, 251–278. Dordrecht: Springer.

Butzer, K.W. 1995. Environmental change in the near east and human impact on the land. In *Civilizations of the ancient near east*, vol. 1, ed. J.M. Sasson, 123–151. New York: Scribner.

Cerling, T.E. 1992. Development of grasslands, savannas in East Africa during the Neogene. *Palaeogeography Palaeoclimatology Palaeoecology* 97: 241–247.

Cerling, T.E., and R.L. Hay. 1988. An isotopic study of paleosol carbonates from Olduvai Gorge. *Quaternary Research* 25: 63–78.

deMenocal, P.B. 1995. Plio-Pleistocene African climate. *Science* 270: 53–59.

deMenocal, P.B. 2004. African climate change and faunal evolution during the Pliocene-Pleistocene. *Earth and Planetary Science Letters* 220: 3–24.

Fenton, M., S. Geiselhart, E.J. Rohling, and Ch Hemleben. 2000. Aplanktonic zones in the Red Sea. *Marine Micropaleontology* 40: 277–294.

Fleitmann, D., and A. Matter. 2009. The speleothem record of climate variability in Southern Arabia. *Comptes Rendus Geoscience* 341: 633–642.

Fleitmann, D., S.J. Burns, A. Mangini, M. Mudelsee, J. Kramers, I. Villa, U. Neff, A.A. Al-Subbary, A. Buettner, D. Hippler, and A. Matter. 2007. Holocene ITCZ and Indian monsoon dynamics recorded in stalagmites from Oman and Yemen (Socotra). *Quaternary Science Reviews* 26: 170–188.

Fleitmann, D., S.J. Burns, M. Pekala, A. Mangini, A. Al-Subbary, M. Al-Aowah, J. Kramers, and A. Matter. 2011. Holocene and Pleistocene pluvial periods in Yemen, southern Arabia. *Quaternary Science Reviews* 30: 783–787.

Geological Society of Oman. 2006. *Oman's geological heritage*, 247. Petroleum Development Oman, Muscat: Sultanate of Oman.

Glennie, K.W., and A.K. Singhvi. 2002. Event stratigraphy, paleoenvironment and chronology of SE Arabian deserts. *Quaternary Science Reviews* 21: 853–869.

Leuschner, D.C., and F. Sirocko. 2000. The low latitude monsoon climate during Dansgaard-Oeschger cycles and Heinrich events. *Quaternary Science Reviews* 19: 243–254.

Leuschner, D.C., and F. Sirocko. 2003. Orbital insolation forcing of the Indian monsoon – A motor for global climate changes? *Palaeogeography, Palaeoclimatology, Palaeoecolgy* 197: 83–95.

Lieberman, D.E. 2001. Another face in our family tree. *Nature* 410: 419–420.

Mix, A.C., N.G. Pisias, W. Rugh, J. Wilson, A. Morey, and T.K. Hagelberg. 1995. Benthic foraminifer stable isotope record from Site 849 (0–5 Ma): Local and global climate changes. In *Proceedings ODP scientific results*, ed. N.G. Pisias, L.A. Mayer, T.R. Janecek, A. Palmer-Julson and T.H. Van Andel, 371–412.

Parker, A.G. 2008. Pleistocene climate change in Arabia: Developing a framework for hominid dispersal over the last 350 ka. In *The evolution of human populations in Arabia*, ed. M.D. Petraglia and J.I. Rose, 39–49. Dordrecht: Springer.

Petit-Marie, N., P. Carbonel, J.L. Reyss, P. Sanlaville, A. Abed, R. Bourroulh, M. Fortugne, and S. Yasin. 2010. A vast Eemian paleolake in southern Jordan (29N). *Global and Planetary Change* 72: 368–373.

Pollock, S. 1999. *Ancient Mesopotamia*, 259. Cambridge: Cambridge University Press.

Preusser, F. 2009. Chronology of the impact of quaternary climate change on continental environments in the Arabian peninsula. *Comptes Rendus Geosciences* 341: 621–632.

Preusser, F., D. Radies, and A. Matter. 2002. A 160,000-year record of dune development and atmospheric circulation in southern Arabia. *Science* 296: 2018–2020.

Preusser, F., D. Radies, F. Driehorst, and A. Matter. 2005. Late quaternary history of the coastal Wahiba Sands, Sultanate of Oman. *Journal of Quaternary Science* 20: 395–405.

Radies, D., F. Preusser, A. Matter, and M. Mange. 2004. Eustatic and climatic controls on the development of the Wahiba Sand Sea, Sultanate of Oman. *Sedimentology* 51: 1359–1385.
Radies, D., S.T. Hasiotis, F. Preusser, E. Neubert, and A. Matter. 2005. Paleoclimatic significance of early Holocene faunal assemblages in wet interdune deposits of the Wahiba Sans Sea, Sultanate of Oman. *Journal of Arid Environments* 62: 109–125.
Rose, J.I., and V.I. Usik. 2009. The "upper paleolithic" of south Arabia. In *The evolution of human populations in Arabia*, ed. M.D. Petraglia and J.I. Rose, 169–185. Dordrecht: Springer.
Rosenberg, T.M., F. Preusser, D. Fleitmann, A. Schwalb, K. Penkman, T.W. Schmid, M.A. Al-Shanti, K. Kadi, and A. Matter. 2011. Humid periods in southern Arabia: Windows of opportunity for modern human dispersal. *Geology* 39: 1115–1118.
Rosenberg, T.M., F. Preusser, J. Risberg, A. Plikk, K.A. Kadi, A. Matter, and D. Fleitmann. 2013. Middle and Late Pleistocene humid periods recorded in palaeolake deposits of the Nafud desert, Saudi Arabia. *Quaternary Science Reviews* 70: 109–123.
Shackleton, N.J., A. Berger, and W.R. Peltier. 1990. An alternative astronomical calibration of the lower Pleistocene time-scale based on ODP site 677. *Transaction of Royal Society of Edinburgh Earth Science* 81: 251–261.
Vaks, A., M. Bar-Matthews, A. Matthews, A. Ayalon, and A. Frumkin. 2010. Middle-late quaternary paleoclimate of northern margins of the Saharan-Arabian desert: Reconstruction from speleothems of Negev Desert, Israel. *Quaternary Science Reviews* 29: 2647–2662.
Westgate, J.A., P.A.R. Shane, N.J.G. Pearce, W.T. Perkins, R. Korisetter, C.A. Chesner, M.A.J. Williams, and S.K. Acharyya. 1998. All Toba tephra occurrences across peninsular India belong to the 75,000 yr B.P. eruption. *Quaternary Research* 50: 107–112.
Wynn, J.G. 2000. Paleosols, stable carbon isotopes, and paleo-environmental interpretation of sKanapoi, Northern Kenya. *Journal of Human Evolution* 39: 411–432.

# Chapter 5
# Earthquake Activity in West Asia: Seismicity in the Mediterranean Sea and Evaluation of the Strong Motion for the AD 365 Crete Earthquake Using the Stochastic Green's Function

**Tsuneo Ohsumi and Yuji Yagi**

The West Asian region is an active area of crustal deformation where many historically huge earthquakes have occurred and crustal movement has continued up to the present. The severe Crete earthquake of the 4th century produced a huge tsunami that caused heavy damage throughout the Mediterranean region. This paper attempts to reproduce the ground motions of the 365 Crete earthquake by using stochastic Green's function method with realistic phases information from observed waveforms of 2013 Crete Island earthquake (Mw 6.4) and its aftershock.

## 5.1 Introduction

West Asia is an area of active crustal deformation and has a history of large magnitude earthquakes. As crustal movement continues today, investigation of seismicity in this region may contribute to understanding and protecting against earthquake and tsunami disasters.

West Asia is a region with very active crustal movement and many strong earthquakes have occurred there since the start of recorded history. One recent destructive earthquake is the 1999 İzmit earthquake (*M* 7.4) near Istanbul, an interplate

T. Ohsumi (✉)
Integrated Research on Disaster Risk Reduction, University National Research Institute for Earth Science and Disaster Resilience (NIED), Tsukuba, Japan
e-mail: t_ohsumi@bosai.go.jp

Y. Yagi
Faculty of Life and Environmental Sciences, University of Tsukuba, 1-1-1, Tennodai, Tsukuba 305-8572, ibaraki, Japan

 
A. Tsuneki et al. (eds.), *Ancient West Asian Civilization*,
DOI 10.1007/978-981-10-0554-1_5

earthquake that occurred along the North Anatolian Fault. The dead and injured numbered 60,000 and several hundred thousand people lost their homes. Generally, strain builds up an order of magnitude faster along the boundaries of plates than in other areas therefore, earthquakes occur repeatedly. The Bam earthquake (*M* 6.3), which struck the World Heritage Site Arg-é Bam in 2003, was an intraplate earthquake that occurred inside a plate. Arg-é Bam is the site of a former citadel in Iran. The buildings and citadel were built with adobe, a building material made by adding water to sand, clay and organic materials such as straw, and kneading the mixture. Its susceptibility to the shaking from earthquakes worsened the damage. Dwellings and other buildings in West Asia are built with stone and adobe building materials; therefore, they are susceptible to this shaking also. Even mid-sized earthquakes quite often cause severe damage.

During comparison of plate boundaries where earthquakes often occur and night time views from space it is interesting to note that areas with strong light are distributed along plate boundaries. In other words, human activity is distributed along plate boundaries. This suggests areas in West Asia suited to cultivation, where land and water are available, or where it is easy to make roads, are found in the vicinity of tectonic zones and fault zones. Arg-é Bam is a city built around an oasis where water emerges from the ground due to a fault. The North Anatolian Fault comprises not only one plane but a number of them, with basins having formed between fault planes in some places. Erzincan, a city in eastern Turkey, is situated in a basin thus formed. An *M* 7.8 earthquake struck this region in 1939.

Investigating what kind of ground motion struck cities in past earthquakes is important for both mitigating the risk of future earthquakes and understanding what types of earthquakes struck people in ancient times. In modern times, when methods for reproducing the ground motion of severe earthquakes have been established and seismic source models for earthquakes have been built, it has become possible to simulate the strong ground motion of various kinds of earthquakes. This means that by determining the seismic source model for an ancient earthquake, it would be possible to reproduce the kind of tremors that struck people in ancient times. The severe Crete earthquake of the fourth century produced a huge tsunami that caused heavy damage throughout the Mediterranean region. The crustal movement caused by that earthquake is recorded along Crete's coastline, and analyzing that record to find slippage distribution makes it possible to reproduce the Crete earthquake's ground motion. This section attempts to reproduce the ground motion observed by the ancients.

According to Hori and Kaneda (2013), the relative plate motion of tectonics from the area around the Mediterranean Sea and west to India is 2–4 cm/year. This movement is smaller than other convergent plate boundaries that are similar to the collision zone in the Himalayas–Tibet mountain belt. However, they are characterized by a subduction zone (solid line in Fig. 5.1) and a predominance of strike-slip

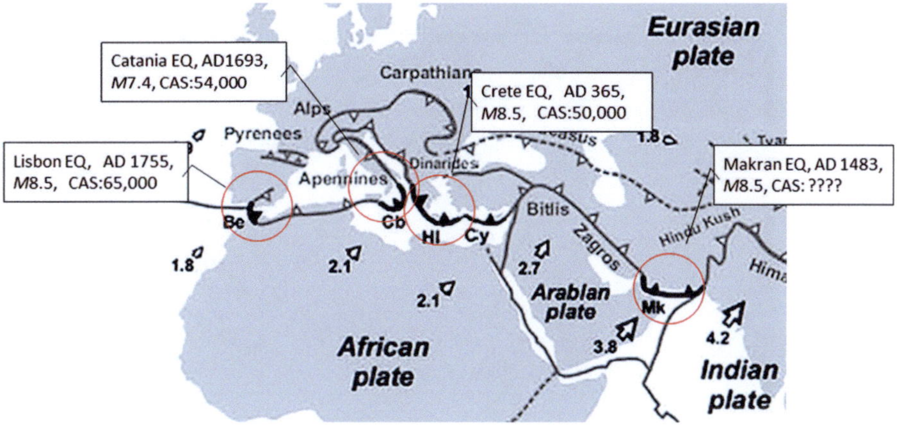


**Fig. 5.1** Tectonics from the area around Mediterranean Sea and that west to India (Modified from Schellart and Rawlinson 2010). Some subduction zones in Mediterranean (Be: Betic—Rif, Cb: Calabria, Hl: Hellenic, Mk: Makran) and North Anatolian fault in Turkey are mentioned (Hori and Kaneda 2013)

fault such as partly exists in the north Anatolia dislocation. These subduction zones cause earthquakes and tsunamis. These areas caused tsunami damage in the Marmara Sea along the north Anatolia fault in the past.

In West Asia, the African plate is being subducted beneath the Anatolian plate at a rate of 1–3.5 cm/year and large magnitude earthquakes frequently occur in this subduction zone. In AD 365 (Fig. 5.2), a large magnitude (*M*8.5) earthquake occurred near Crete (e.g., Fischer (2007), Shaw et al. (2008), Stiros (2010), Papadimitriou and Karakostas (2008)). The AD 365 earthquake, one of the best known ancient earthquakes in the eastern Mediterranean and caused a tsunami that resulted in great damage to Syria, northern Egypt and the Greek coast. According to Pirazzoli (1986), who investigated the upheaval of the coastline in the eastern Mediterranean, the period between 350 and 550 AD was the one of the most seismically active periods in the past 2000 years.

Crete, located 160 km south of the Greek mainland, has the largest (area of 8336 km$^2$) among the approximately 3000 Islands in the Aegean. Ancient earthquakes in Crete have been reported in various books by Ambraseys et al. (e.g., 1994). In the fourth century, Ammiaus, a historian and soldier wrote a series of history books consisting of 31 volumes. Due to Christian propagation, which began in the Roman Empire, many historical records were made in the fourth–fifth centuries.

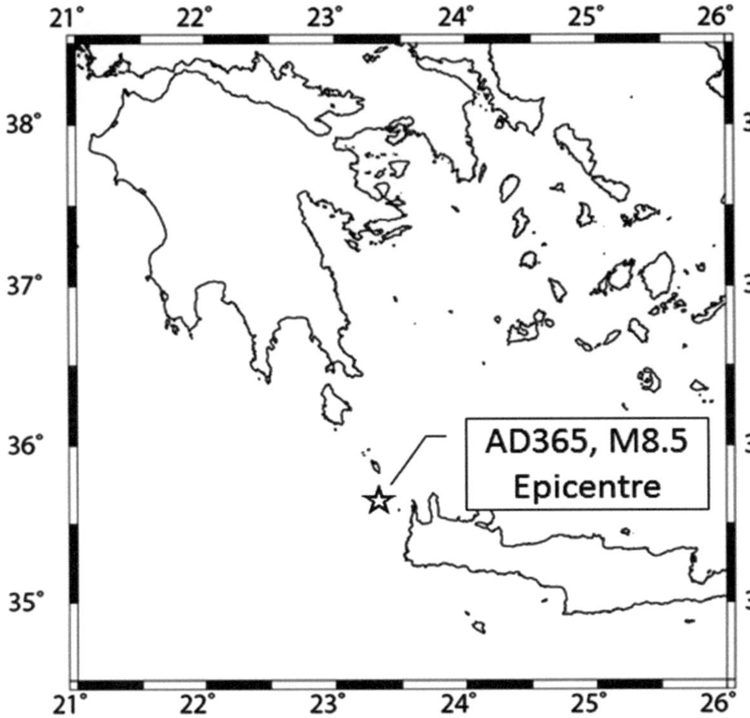


**Fig. 5.2** Epicentre of The AD 365 Crete Earthquake

The tsunami caused by the AD 365 earthquake resulted in more damage in the Peloponnese peninsula than from the earthquake itself. Because the magnitude was higher than M8, tremors were propagated across a large area surrounding the Mediterranean. Smaller magnitude earthquakes that recently occurred in the Greek Islands also affected a wide area around the Mediterranean. John Cassian, a theologian in the fourth–fifth century, and Sozomenes, a Byzantine historian in the fifth century, described evidence of widespread flooding from the tsunami by tracing damage on the roofs of buildings, and subsequent retreat of the coastline. A Byzantine historian, George the Monk, mentioned the tsunami in a ninth century chronicle. The tsunami caused by the AD 365 earthquake was also chronicled by Theopanos in the eighth–ninth centuries, Cedrenus in the eleventh century, and Glycas in the twelfth century. According to the literature, the tsunami destroyed 50,000 houses and caused 5000 casualties in Alexandria, Egypt.

As large magnitude earthquakes and associated disasters could occur again in the future, this paper investigates and describes the characteristics of the AD 365 Crete earthquake.

## 5.2 Mythology of the Earthquake and Tsunami

### *5.2.1 The Presence of Giants*

In Greek mythology, natural disasters were supposed to be caused by the Gigantes' fight against the gods led by Zeus. In this story, Enceladus is one of the Gigantes and causes earthquakes and tsunamis. The goddess Athena, leaped from Zeus's head, fully grown and armed, entered the battle against the Gigantes and defeated Enceladus thus calming the earthquake and tsunami.

The 12 Olympian gods Zeus, Hera, Athena, Apollon, Aphrodite, Ares, Artemis, Demeter, Hephatistos, Hermes, Poseidon and Hephaistos (or Dyonissos) appeared as part of the motif on the pediment of the Pantienon royal palace. Most parts of the pediment were destroyed during the Roman Empire; however, some are stored in the Acropolis Museum (Fig. 5.3) and the British Museum. Replicas of the 12 Olympians are displayed in the Academy of Athens (Fig. 5.4) and in the Akropoli subway station, together with pottery excavated during the construction of the station.

**Fig. 5.3** West side of the Parthenon. Most parts of the pediment were destroyed during the Roman Empire period (Photo by T. Ohsumi)

**Fig. 5.4** Poseidon found in Mykonos with a dolphin, as its symbol, next to the leg. It is said that it originally grasped a trident in its right hand (National Archaeological Museum: photo by T. Ohsumi)

## *5.2.2 Poseidon: God of the Sea*

Poseidon was the God of the sea for the ancient Greeks (Fig. 5.3). Sometimes Poseidon calmed the sea and, at other times, caused storms. He made the sea storms with his trident and caused major earthquakes and tsunamis with his anger and created the topography of the islands in ancient Greece. On the other hand, he sometimes calmed natural disasters. There is a myth that Athena and Poseidon fought over control of Athens. Poseidon damaged the earth by using his trident, while Athena planted an olive. Athena was welcomed by the citizens who named their city after her and built the Parthenon to honor her.

## 5.3 Crustal Movements

Flemming (1978) evaluated land subsidence and upheaval and estimated the relative rate of annual sea level rise (1.05 mm/year.) based on research on the south-west coast of Turkey and at about 175 points in Cyprus. Change in the sea surface in western Crete under the condition of a continuous sea-level rise and intermittent land uplift is shown in Fig. 5.5.

Pirazzoli et al. (1986) suggests that there are traces of upheaval along the Greek coast from an earthquake that occurred between the mid-4th and mid-6th centuries in the Early Byzantine tectonic paroxysm (EBTP) turbulent period. Pirazzoli et al. revised his previous interpretation (1982; Pirazzoli 1986) after detailed survey and radiocarbon dating of the samples obtained from Antikyhira Island. Significant co-seismic uplift that took place during a short period was demonstrated by over 30 radiocarbon dates from 12 regions in Greece and by very precise sea-level indicators in the eastern Mediterranean. Therefore, it is assumed that the scale of uplift in Crete was 0.5–1.0 m in general but gradually increased towards the south-west and reached approximately 9 m. Radiocarbon dates show that the largest change occurred between AD 261 and AD 425.

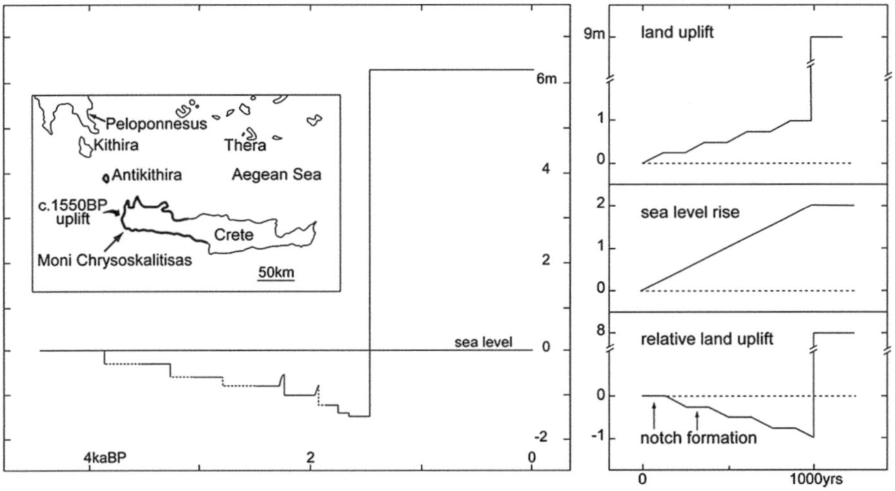


**Fig. 5.5** *Left*: Upheaval history of the oldest (lowermost) layer, assuming a fixed sea-level based on the data by Thommeret et al. (1981) and Pirazzoli et al. (1982). *Right*: A scenario explaining the change in the relative sea level in west Crete based on the hypothesis of constant rise of the sea level and intermittent rise of the land (*after* Stiros 1996)

## 5.4 Seismic Activity

### 5.4.1 Ancient Earthquake Damage

Minoan Knossos is located in Iraklio, the administrative centre of Crete. Minoan Knossos reached its heyday in the eighteenth–sixteenth centuries BC and suddenly disappeared in the fifteenth century BC.

#### 5.4.1.1 Aptra

Aptra is an ancient city located 30 km east of Chania and 120 km west of Iraklio (Fig. 5.6). It was built in the fifteenth to fourteenth centuries BC and was severely damaged by huge earthquakes in the fourth and seventh centuries. Roman occupation of Aptra began in 69 BC, as evidenced ancient temple ruins, a castle gate, and walls. Catastrophic earthquake damage occurred in 365, concurrent with the fall of the Roman Empire.

#### 5.4.1.2 Seismic Intensity Distribution of Past Earthquake

Figure 5.7 shows the distribution of seismic intensity for the ancient earthquakes (1886, 1903, 1926 & 1956) around Crete (Sieberg 1932). According to Wyss and Baer (1981), the characteristics of these earthquakes may be summarized as below.

1. Rupture affected the entire Hellenic arc.
2. Earth tremor was felt over a vast area.
3. Isoseismal contours are asymmetric. Fairly high intensities are observed in areas distant from the Hellenic arc, while intensities at the other side of the arc decreased sharply.

**Fig. 5.6** *Left*: Hypothetical representation of the fortification tower and entrance (from Information plate of National Strategic Reference Framework (NSRF) 25th Department of Antiquities). *Right*: Entrance of the castle gate of Aptra (Photo by T. Ohsumi)

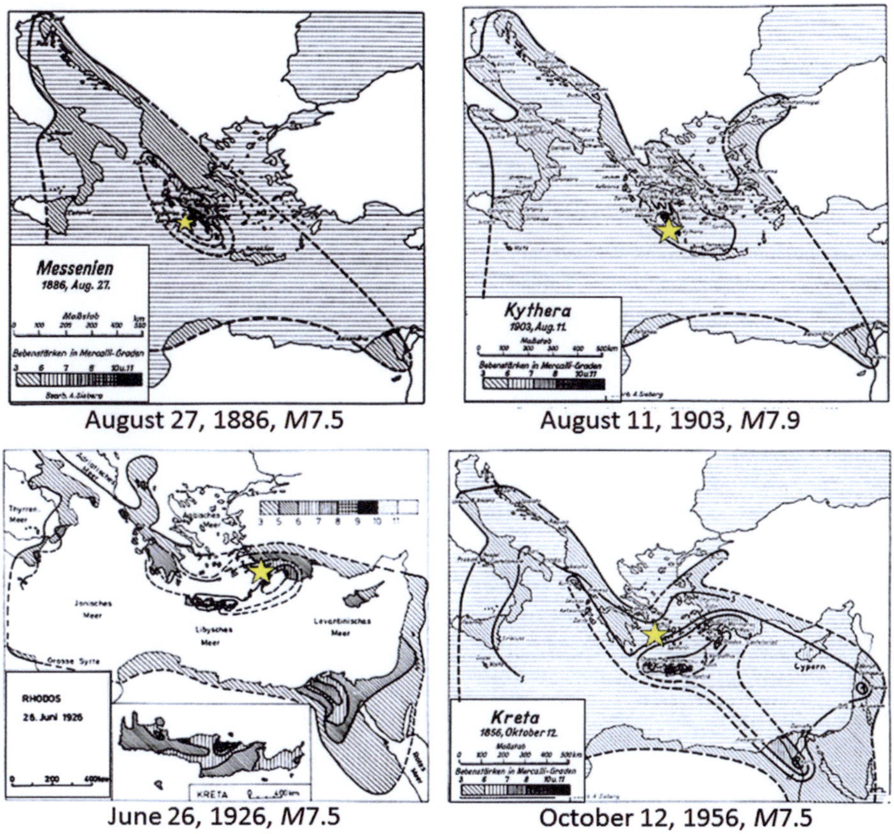


**Fig. 5.7** The distribution of the seismic intensity of the ancient earthquakes (1886, 1903, 1926, 1956) around Crete (Sieberg 1932)

4. Epicentres of these earthquakes are located offshore south of the Island arc and the assumed intensities of all these earthquakes reaching a degree of XI.

#### 5.4.1.3 Seismic Hazard Map

Sieberg (1932) created a seismic map of the Levant. Figure 5.8 shows the Crete region, with extensive damage to buildings shown in red. In the 1926 earthquake (*M*7.5), the area stretching from Iraklio to Knossos recorded modified Mercalli seismic intensity of 9–10. Chania in the north-west of Crete, Sitia in the north-east, Ierapetra in the south-east and Psiloritis in the Ida Mountain areas were expected to be high seismic intensity areas.

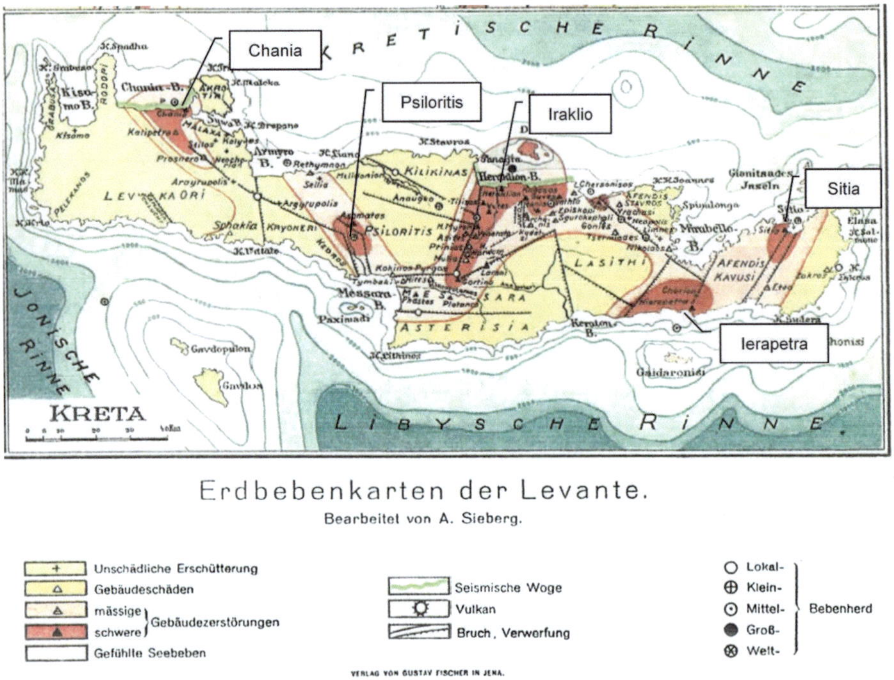


**Fig. 5.8** Earthquake map of the Crete Island (Modified from Sieberg 1932)

#### 5.4.1.4 Earthquake Scale

Stiros (2010) conducted an elastic inversion analysis of the coastal upheaval and suggested that this earthquake had a minimum magnitude of 8.5 and was due to the reverse fault located offshore of south-western Crete. Shaw et al. (2008) showed a fault model that accounted for upheaval on Crete. Based on an investigation of the strike of the Aegean arc and the depth of epicentres for small earthquakes, Stiros (2010) suggests that the depth of the fault is shallower than 70 km.

## 5.5 Estimation of Earthquake Ground Motions

In this study, we estimate the earthquake ground motions by the stochastic Green's function method. Table 5.1 shows parameters by each researcher. Fischer (2007) set dip as 13° and a low angle along a subduction plate, and set dislocation to 42 m. Shaw et al. (2008), Stiros (2010) and Papadimitriou and Karakosta (2008) suggested for the parameters set that the shallow branch of the subduction zone dips at low angle to and couples with the Aegean lithosphere, while the deep branch dips freely (without coupling) at a high angle beneath the south Aegean trough.

**Table 5.1** Parameters setting

| | | Fischer (2007) | Shaw et al. (2008) | Stiros (2010) | Papadimitriou and Karakostas (2008) |
|---|---|---|---|---|---|
| Strike | | 297 | 315 | 292.5 | 315 |
| Dip | | 13 | 30 ± 5 | 40 | 35 |
| Depth | km | | 45 | 70 | 5–50 |
| Length | km | 145 | 100 | 105 | 160 |
| Width | km | 130 | | 100 | 80 |
| Slip | m | 42 | 20 | 16 | 8.9 |
| $M_0$ | dyne cm | | | $5.04\times10^{28}$ | $5.7\times10^{28}$ |
| $M_w$ | | 8.5 | 8.3–8.5 | 8.5 | 8.3 |

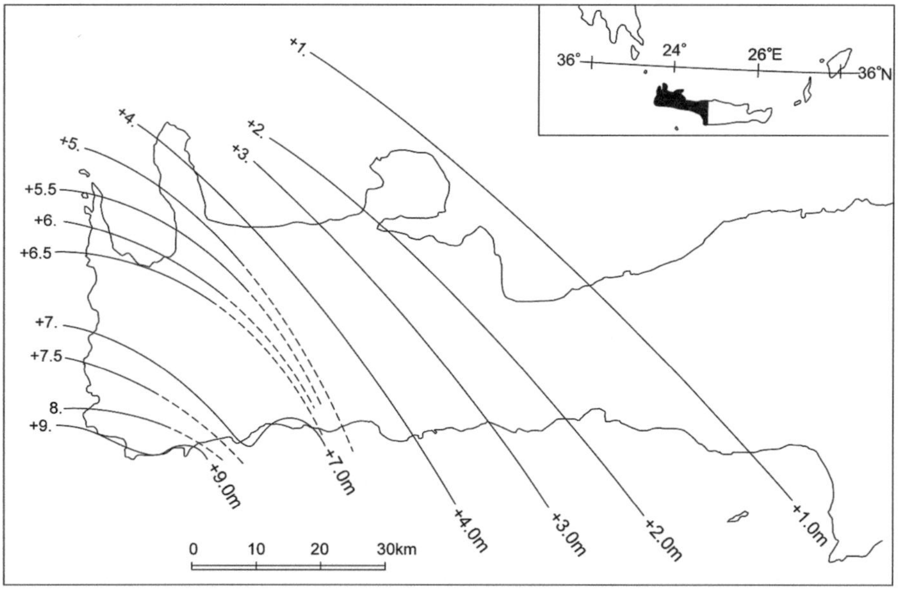


**Fig. 5.9** Contour of upheaval Crete (up to 9 m of the island southwest) (Modified from Pirazzoli et al. (1996))

### *5.5.1 Parameter Setting*

Pirazzoli et al. (1996) suggests that there are traces of upheaval along the Crete coast from radiocarbon dating and a detailed survey indicated evidence of Holocene coseismic activity (Fig. 5.9). By using this data, comparison of uplift distribution was performed with these parameters. According to Murotani et al. (2013), the average dislocation was around 10 m for the 1960 Chile earthquake and the 2011 Tohoku earthquake. The average dislocation of 42 m was four times the value of the 1960 Chili earthquake and the 2011 Tohoku earthquake. Moreover, sufficient value of the

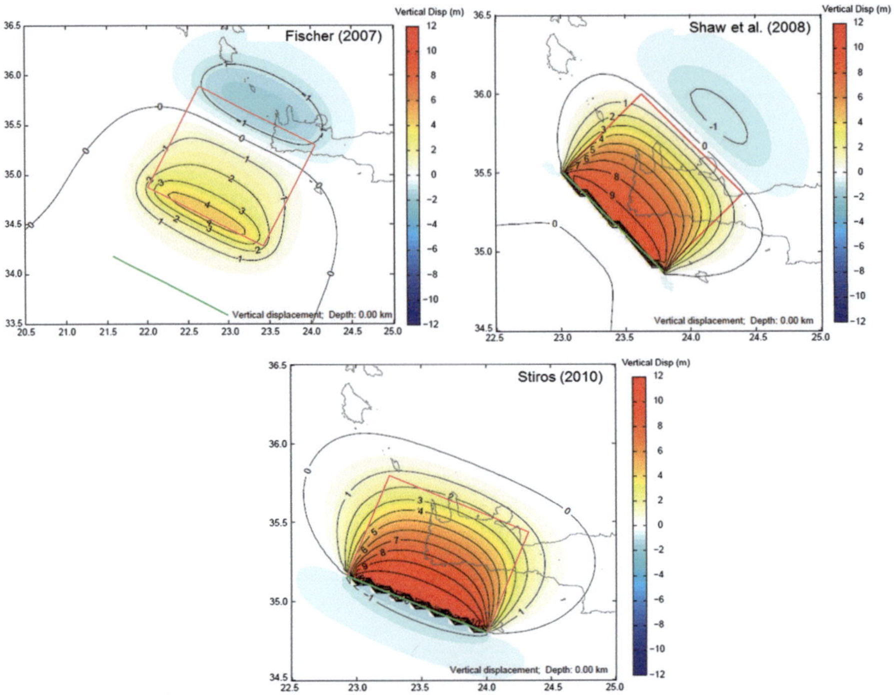


**Fig. 5.10** Crustal displacements, upheaval generating areas by Fischer (2007), Shaw et al. (2008) and Stiros (2010) models

upheaval distribution is not represented through Okada's formula (1992) (Fig. 5.10). Therefore it is difficult to explain the uplift distribution using the shallow dip angle model.

Shaw et al. (2008) shows the seismicity and topography in the area of Crete with the seismicity of the region corresponding to the AD 365 earthquake. According to Shaw et al. (2008), distribution of the upheaval suggested that the AD 365 Crete earthquake occurred not on the subduction interface beneath Crete, but on a fault dipping at about 30° within the overriding plate. The shallow branch of the subduction zone dips at low angle to and couples with the Aegean lithosphere, while the deep branch dips freely (without coupling) at a high angle beneath the south Aegean trough. Therefore we apply the crustal upheaval modeling study by Shaw et al. (2008) using the formula from Okada (1992). Shaw et al. (2008) set dip angle at 30° and a high angle within the overriding plate, the strike was set at 315°, dislocation at 20 m and fault depth at 45 km. The strike had the same angle of 315° as the fault plain solution from Papadimitriou and Karakosta (2008).

Stiros (2010) set a steep dip angle of 40°; strike angle of 292.5° and a uniform dislocation of 16 m. Stiros (2010) shows that elastic dislocation analysis of coastal

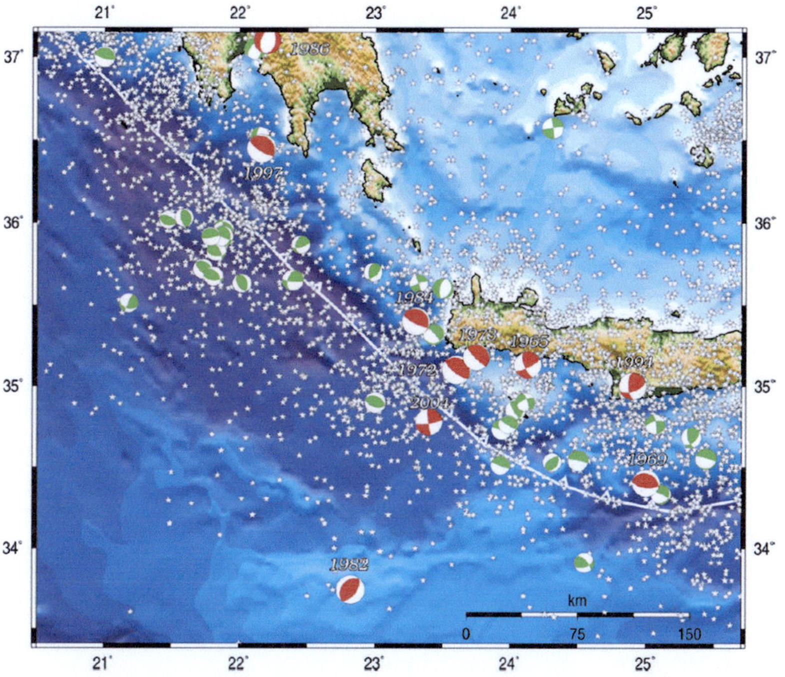


**Fig. 5.11** Seismicity and fault plane solutions as lower-hemisphere equal-area projections along the southwestern part of the Hellenic Arc. Beach balls in *red* represent events with M ≥ 6.0 while *green* ones the smaller magnitude events

upheaval data reveals that this earthquake was associated with a reverse fault offshore of southwestern Crete, and that its minimum magnitude was 8.5. This model was consistent with the approximate seabed trace of the fault, which observed and calculated displacements due to the modelled fault, and fault depth was as deep as 70 km. We apply the crustal upheaval modelling from Stiros (2010) using the formula of Okada (1992).

Papadimitriou and Karakostas (2008) set a direction of 315 based on Papazachos et al. (2000, 2001). The earthquake moment assumed $5.7 \times 10^{28}$ dyne cm from the area of the dislocation and an elastic coefficient.

According to Papadimitriou and Karakostas (2008), seismic coupling has been correlated with the maximum size of earthquakes that occur at a subduction zone. Subduction zones that are strongly coupled seismically, periodically produce great earthquakes ($Mw > 8.0$) (Kanamori 1977), while those that are seismically uncoupled produce only moderate to large earthquakes ($Mw < 8.0$) (Ruff and Kanamori 1980) (Fig. 5.11). Reverse faulting is observed on planes with a NW or NE strike, and with approximately E-W P axis, the larger of them occurring in 1982 beneath the Mediterranean ridge. The slip distribution calculated by Papadimitriou and

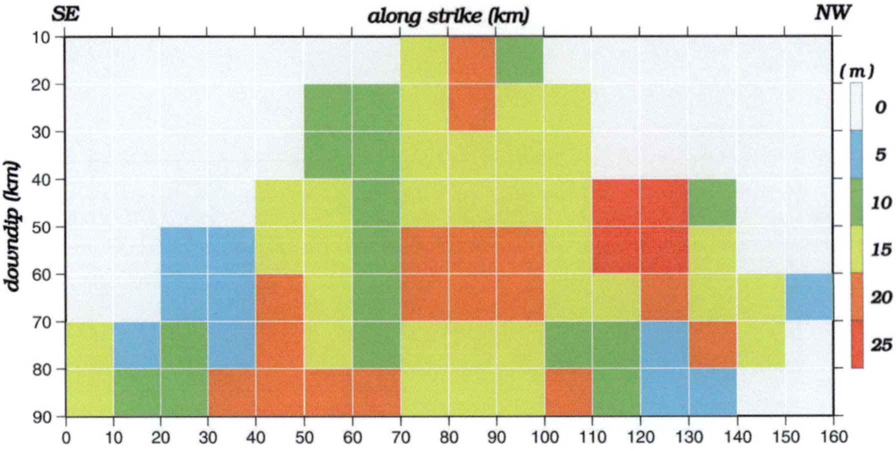


**Fig. 5.12** Characteristics of slip distribution of $M \sim 9$ earthquakes in the world Scaling relations of seismic moment, rupture area, average slip, and asperity size for *M*9 subduction-zone earthquakes (*after* Papadimitriou and Karakostas 2008)

Karakosta (2008) indicated a non-uniform distribution in the fault plane (Fig. 5.12). This study caused upheaval by using each of the 128 mesh divisions to give the dislocation. In addition, the geometry of the fault was almost the same as the setting of Shaw et al. (2008), except the fault length was set at 160 km. Shaw et al. (2008) had a fault length of 100 square km, a *Mw* scale of 8.0 through using a low formulae scaling of the earthquake size and fault area (e.g. R. Sato: log $S = M–4.07$). The *M* 8 level is the difference of 8.3–8.5 defined by each researcher. Thus, ground motion was estimated by the stochastic Green's function method using the parameters of Papadimitriou and Karakostas (2008).

### 5.5.2 Small Event

The stochastic Green's function uses random phases or realistic phases. This study used realistic phases scrutinized from earthquake records. Earthquake records of Crete Island, *Mw*6.4, 12 Oct. 2013, give these phases for the stochastic Green's function. The dislocation model is based on Papadimitriou and Karakosta (2008). Some of these velocity records saturated (over scale) at Antikythera, Chana and Iraklio (Fig. 5.13 upper graph). Considering the scaling relationship, *Mw*6.4 is a better selection as small event although another selection is used in smaller magnitude records. Thus, we selected the screening data for the aftershock record (*Mw*4.0, 12 Oct. 2013: Fig. 5.13 lower graph) at Antikythera, Chana, Aptera and Iraklio. At Phalasarna, there is no recording point. Thus, the Aptra record is used for the estimation at Phalasarna. At Athens, the aftershock record included noise therefore we used the main shock phase at this point.

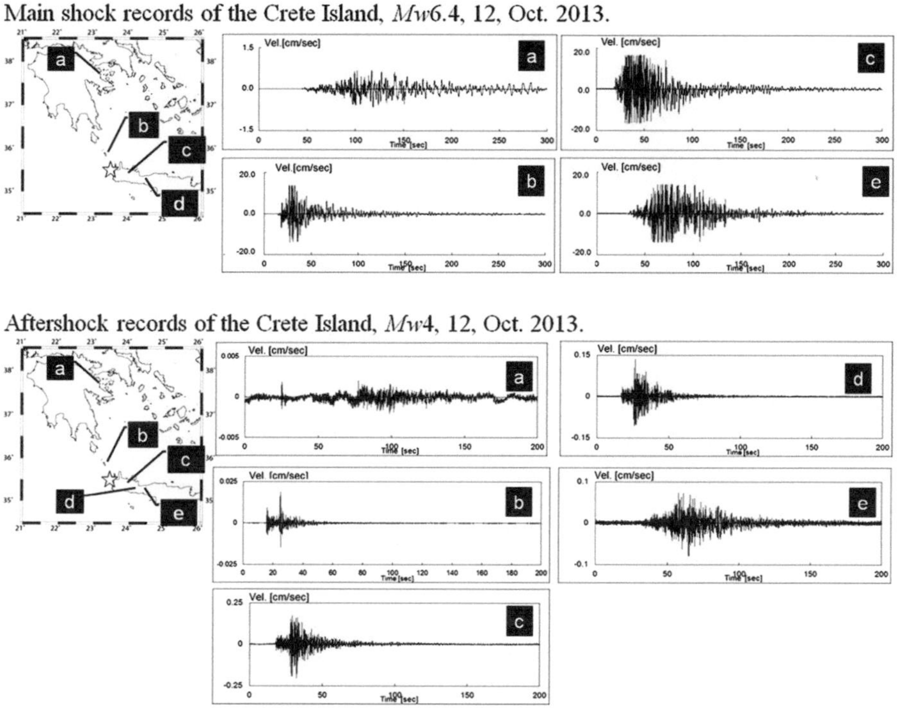


**Fig. 5.13** Earthquake records of the Crete Island

### 5.5.3 *Site-Specific Data*

The vertical shear velocity is the Vs value in the north-west area defined by Karagianni (2005). The Vs value was set at a typical value from Athens, Antikythera and Aptera as Vs 3.5 km/s is 5 km from 2.5 km depth and Vs 1.5 km/s is shallower than 2.5 km depth, for site-specific implications. Subsurface structures were referred as deep Vs values provided from Karagianni (2005). Site amplification characteristics of the surface layer ground configuration were chosen to fit the transfer function of the one-dimensional wave theory. GL-100–200 m could be Vs = 700 m/s among surface 200 m.

At Chania with thick sedimentary layers, and Iraklio, GL-0–100 m calculated the H/V spectrum from the portion of Coda waves of a seismic observation result. The result indicates the configuration predominant frequency of 100 m from GL-0 m. The H/V-ratio measurement has become a powerful tool for engineers in estimating ground motion characteristics such as amplification of soil deposits by horizontal/vertical (H/V)-ratio.

The most efficient way to determine the dynamic behavior of structures is to use the H/V-ratio technique. Spectral analyses were performed for measured data at different down-hole locations. It can be seen that the horizontal motions of the surface

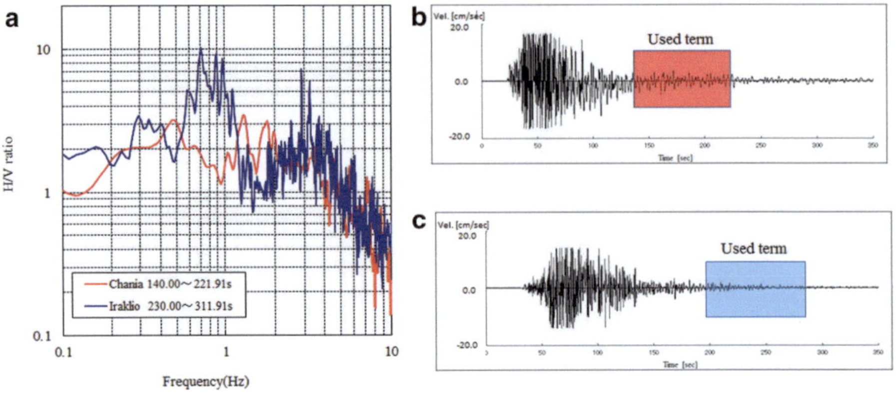


**Fig. 5.14** H/V Spectrum at Chania and Iraklio site. At Chania with thick sedimentary layers, and Iraklio,GL-0–100 m calculated the H/V spectrum (**a**) from the portion of Coda waves (Chania: 140.00–221.91 s (**b**), Iraklio: 230.00–311.91 s (**c**)) of a seismic observation result

waves are very similar to the bottom waves. Result of the calculation indicates (Fig. 5.14), the H/V spectrum peak appeared 0.4 Hz and 1.2 Hz in Chania (the configuration layer was set at a depth of 100 m for seismic response analyses of horizontally layered one-dimensional soil deposits corresponding to the H/V spectrum). The H/V spectrum has two peaks at 3 Hz and 0.7 Hz at Iraklio, which correspond to dominant periods of the 100 m surface layer.

### 5.5.4 *Estimation of the AD 365 Earthquake Ground Motions Waveforms*

Figure 5.15 shows the estimation of the AD 365 earthquake ground motions. At Phalasarna, 10 km from the epicentre, estimated velocity is 102 cm/s. At Antikythera Island, 50 km from the epicentre, the estimated velocity is 57 cm/s. At Chania, 50 km from the epicentre, the estimated velocity is 108 cm/s. At Aptra, 75 km from the epicentre, the estimated velocity is 55 cm/s. At Iraklio, 150 km from the epicentre, the estimated velocity is 41 cm/s. At Athene, 270 km from the epicentre, the estimated velocity is 2 cm/s, these values reduce by distance from the epicentre although the duration time expands to 200 s.

Evidence of a 9 m upheaval by the AD 365 earthquake movement are indicated by first: the detailed survey and radiocarbon dates by Pirazzoli et al. (1996), second: the date of casualties' from coins found in a collapsed house by Stiros (2010), third: the ruins of an old harbor and it's quay wall. Moreover, synthetic upheaval distribution is derived from calculations using the Stiros (2010), Shaw et al. (2008) and Papadimitriou and Karakosta (2008) models.

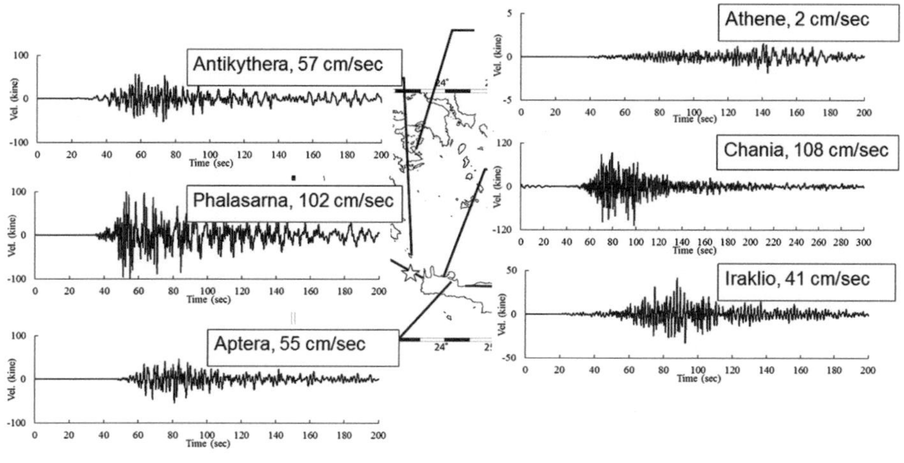


**Fig. 5.15** Estimation of the AD 365 Earthquake synthetic velocity waveforms

Synthetics upheaval distributions with the low angle fault model (e.g. Pirazzoli et al. 1996) and the high angle fault models (e.g. Shaw et al. 2008) were calculated with formula of Okada (1992). The synthetics with the low angle fault model corresponding to inter-plate earthquake is small amplitude in comparison with observation. The observed upheaval distribution can reproduce by the high angle fault models corresponding to intra-plate earthquake in the overriding plate (Shaw et al. 2008; Papadimitriou and Karakost 2008; Stiros 2010).

Stiros (2010) had set dip angle at 40° and a high angle within the overriding plate, the strike angle was 292.5° and dislocation to 16 m. Stiros (2010) saw that elastic dislocation analysis of coastal upheaval data revealed that this earthquake was associated with a reverse fault offshore of southwestern Crete, therefore its minimum magnitude was 8.5. This model was consistent with the approximate seabed trace of the fault, which observed and calculated displacements due to the modelled fault, and fault depth was as deep as 70 km. We applied the crustal upheaval by Stiros (2010) modeling study using the calculation formula of Okada (1992).

Shaw et al. (2008) suggested that the same process takes place along the entire Hellenic subduction zone, and such events may occur approximately once every 800 years. The stochastic Green's function uses random phases or realistic phases. In this study, we use realistic phases from earthquake records. First, we used the earthquake records of the Crete Island, *Mw*6.4, 12 Oct. 2013, and used these phases as the stochastic Green's function. Some of these velocity records were saturated (over scale) at Antikythera, Chana and Iraklio. Since the data is saturated at several stations, we selected smaller magnitude records.

We consider that the scaling relationship, *Mw* 6.4 is a better selection for a small event. Although, we used *Mw* 4, 12 Oct. 2013, the aftershock of 12 Oct. 2013 was *Mw*6.4. Thus, the record of the aftershock of *Mw* 6.4, 12 Oct. 2013 is used as the small event except for Athens.

The Earthquake map of Crete Island shows high risk areas in Chania and Iraklio. Comparison with estimated velocity and this map, indicates high level velocity occurs in Phalasarna (102 cm/s) and Chania (115 cm/s). These velocity levels correspond to the Great Hanshin-Awaji Earthquake, Japan (1995). At Irakulio, even 150 km from the epicentre, the estimated velocity and velocity is 41 cm/s, and duration time is 100 s. This velocity level and duration time will cause liquefaction over a wide area.

At Aptra, 75 km from the epicentre, estimated velocity is 55 cm/s and at this level will cause heavy damage to masonry structure. At Aptra, the fortification tower and the entrance collapsed in the AD 365 earthquake.

## 5.6 Summary and Conclusions

By using the AD 365 earthquake parameters obtained by Shaw et al. (2008) and Papadimitriou and Karakostas (2008), we calculated the earthquake's motions. Observed phase records at seismometer sites located on Crete Island are estimated by using the Statistical Green's function methods for the AD 365 earthquake. Very good results were obtained by applying three Green's functions to the fault geometry over a range of 160 km, where the shallow branch of the subduction zone dips at low angle to and couples with the Aegean lithosphere. Thus, in this stochastic simulation we applied Papadimitriou and Karakostas (2008)'s parameters. Their study used each of the 128 mesh divisions to indicate this dislocation and upheaval. The stochastic Green's function uses these slip distributions at each of the 128 divisions.

The comparison of the velocity response spectrum for the synthetic waveforms at Chania and Irakio sites show dominant natural periods at 0.5 s (2 Hz) and 1.0 s (1 Hz), these periods will cause a non-liner effect in the sedimentary layer. Chania and Iraklio sites represent use of the H/V ratio. This H/V ratio is an effective method for sites with less ground information. In regard to the sediment layer, the synthetic velocity of the Chania site, which is 50 km from the epicentre, is larger than the Phalasarna site velocity, which is 10 km from the epicentre. The velocity response spectrum for the synthetic waveforms at Chania and Irakio sites show the dominant natural period at 2 s, indicating these periods caused a non-liner effect for the sedimentary layer.

**Acknowledgments** The authors acknowledge the kind help during their reconnaissance given by Professor Gerassimos A. Papadopoulos, Institute of Geodynamics National Observatory of Athens, who provided the authors with important shear velocity structures in the Aegean area. The authors are indebted to Professor Eleftheria Papadimitriou and Associate Professor Vassilis Karakostas, Aristotele University of Thessaloniki, who provided the authors with important waveform data obtained from Crete Island.

The authors were provided with important information and constructive comments for the stochastic Green's function method by Mr. Yasuhiro Fukushima, Eight-Japan Engineering Consultants Inc.

## References


Ambraseys, N.N., C.P. Melville, and R.D. Adams. 1994. *The seismicity of Egypt, Arabia and the Red Sea*. Cambridge, MA: Cambridge University Press.

Fischer, K.D. 2007. Modelling the 365 AD Crete earthquake and its tsunami. *Geophysical Research Abstracts* 9, 09458.

Flemming, N. C. 1978. Holocene eustatic changes and coastal tectonics in the northeast Mediterranean: Implications for models of crustal consumption. *Philosophical Transaction of the Royal Society London, Series A* 289 (1362): 405–458 + Appendix I.

Hori, T., and Y. Kaneda. 2013. Giant earthquakes and tsunamis in the world: Mediterranean Sea. *Report of CCEP* 89.

Kanamori, H. 1977. The energy release in great earthquakes. *Journal of Geophysical Research* 82: 2981–2987.

Karagianni, E.E. 2005. Shear velocity structure in the Aegean area obtained by inversion of Rayleigh waves. *Geophysical Journal International* 160(1): 127–143.

Murotani, S., K. Satake, and Y. Fujii. 2013. Scaling relations of seismic moment, rupture area, average slip, and asperity size for M ~ 9 subduction-zone earthquakes. *Geophysical Reserarch Letters* 40(19): 5070–5074.

Okada, Y. 1992. Internal deformation due to shear and tensile faults in a half-space. *Bulletin of the Seismological Society of America* 82: 1018–1040.

Papadimitriou, E., and V. Karakostas. 2008. Rupture model of the great AD 365 Crete earthquake in the south-western part of the Hellenic Arc. *Acta Geophysica* 56(2): 293–312.

Papazachos, B.C., B.G. Karakostas, C.B. Papazachos, and E.M. Scordilis. 2000. The geometry of the Benioff zone and lithospheric kinematics in the Hellenic arc. *Tectonophysics* 319: 275–300.

Papazachos, B.C., D.M. Mountrakis, C.B. Papazachos, M.D. Tranos, G.F. Karakaisis, and A.S. Savvaidis. 2001. The faults that caused the known strong earthquakes in Greece and surrounding areas during 5th century B.C. up to present. In *Proceedings of the 2nd Conference on Earthquake Engineering and Engineering Seismic*, 28–30 September 2001, Thessaloniki 1, 17–26

Pirazzoli, P.A. 1986. The early Byzantine tectonic paroxysm. Z. Geomorph. N.F., Suppl.-Bd 62: 31–49.

Pirazzoli, P.A., J. Thommeret, Y. Thommeret, J. Laborel, and L.F. Montaggioni. 1982. Crustal block movements from holocene shorelines: Crete and Antikythira (Greece). *Tectonophysics* 86: 27–43.

Pirazzoli, P.A., J. Laborel, and S.C. Stiros. 1996. Earthquake clustering in the eastern Mediterranean during historical times. *Journal of Geophysical Research* 101(B3): 6083–6097, Solid earth.

Ruff, L., and H. Kanamori. 1980. Seismicity and the subduction process. *Physics of the Earth and Planetary Interiors* 23: 240–252.

Schellart, W.P., and N. Rawlinson. 2010. Convergent plate margin dynamics: New perspectives from structural geology, geophysics and geodynamic modelling. *Tectonophysics* 483: 4–19.

Shaw, B., N. Ambraseys, P.C. England, M.A. Floyd, G.J. Gorman, T.F.G. Higham, J.A. Jackson, J.-M. Nocquet, C.C. Pain, and M.D. Piggott. 2008. Eastern Mediterranean tectonics and tsunami hazard inferred from the AD365 earthquake. *Nature Geoscience* 1: 268–276.

Sieberg, A. 1932. Isoseismal contours crete ad 365 crete earthquake intensity isoseismal contours crete, Untersuchungen uber Erdbeben und Bruchschollenbau im Oestlichen Mittelmeergebiet, Jena: [s.n.], BA49737430.

Stiros, S.C. 1996. Late holocene relative sea level changes in SW Crete: Evidence of an unusual earthquake cycle. *Annali di Geofisica* 39(3): 677–687.

Stiros, S.C. 2010. The 8.5+ magnitude, AD365 earthquake in Crete: Coastal uplift, topography changes, archaeological and historical signature. *Quaternary International* 216: 54–63.

Thommeret, Y., J. Thommeret, P.A. Pirazzoli, L.F. Montaggioni, and J. Laborel. 1981. Nouvelles donnees sur les rivages souleves de I'Holocene dans l'ouest de la Crete. *Oceanis* 7(4): 473–480.

Wyss, M., and M. Baer. 1981. Earthquake hazard in the Hellenic Arc. *Maurice Ewing Series* 4: 153–172.

# Part II
# Great Transformations in Prehistory

# Chapter 6
# The Origins of Agriculture

**Ken-ichi Tanno and Osamu Maeda**

Today, cereals and legumes constitute a major part of the diet that we consume every day. Many foods that we eat on a daily basis comprise wheat, barley and other cereals as well as various peas and beans. However, these crops were not a major food resource for the most part of human history and became a daily food source only after they began to be cultivated about 10,000–12,000 years ago. This means that the first domestication of these crops, that is, the origin of agriculture, was an epoch-making episode, which underpinned the foundation of modern civilization.

West Asia is known to be the place of the origin of agriculture. The earliest evidence of plant cultivation in human history is known from this region where various crops, such as wheat, barley, lentils and chickpeas, were first domesticated.[1] In archaeology, the process through which the first agricultural practice emerged and developed is an important area of study and has recently become a contentious issue of debate, as several new hypotheses have been proposed. This chapter will review past studies on the origin of agriculture. First, the nuclear zone hypothesis and a climate-driven model will be briefly reviewed. Then, a critical overview will be provided for the "core area hypothesis", which has been suggested based mainly on the geographical distributions of wild progenitors of domestic crops and the results

[1] The term "cultivation" refers to a human activity involved in managing plant resources, such as clearing, plowing, sowing, planting or harvesting, regardless of the morphologically wild or domestic status of the plants. "Domestication" refers to genetic and morphological changes in plant evolution from wild to domestic status resulting from selection under human influence through cultivation. Domestic-type plants have traits such as reduction in natural seed dispersal, larger seed/fruit size or loss of germination inhibition.

K.-i. Tanno (✉)
Faculty of Agriculture, Yamaguchi University,
1677-1, Oaza Yoshida, Yamaguchi 753-8515, Yamaguchi, Japan
e-mail: tanno@yamaguchi-u.ac.jp

O. Maeda
Institute for Comparative Research in Human and Social Sciences, University of Tsukuba,
1-1-1, Tennodai, Tsukuba 305-8571, Ibaraki, Japan



A. Tsuneki et al. (eds.), *Ancient West Asian Civilization*,
DOI 10.1007/978-981-10-0554-1_6

of a DNA study of einkorn wheat. Finally, an alternative view on the origin of agriculture based on a careful examination of the latest archaeobotanical evidence is presented.

## 6.1 Changing Perspectives on the Origin of Agriculture

In 1948, the Iraq-Jarmo project led by Robert Braidwood began in the foothills of Central Zagros in Iraqi Kurdistan (Braidwood and Howe 1960; Braidwood et al. 1983). The project was the first archaeological research to have adopted a multidisciplinary approach, involving archaeobotanical study in an attempt to investigate the origin of the domestication of plants and animals. In the 1950s, a series of excavations were performed at the site of Jarmo, revealing that the site was a Neolithic farming village where wheat (emmer and einkorn) and barley were cultivated. Dated to around the 8th millennium cal. BC by the radiometric dating method that was newly invented at that time, it was believed that the ancient village of Jarmo was the oldest agricultural settlement in the world. This theory was widely accepted by both academics and the general public and appeared, for example, in world history textbooks in Japanese schools. Braidwood and his colleagues argued that the agricultural origin must be found in the habitats of wild cereals and legumes, that is, the nuclear zone, which extends across the hilly regions adjacent to the Fertile Crescent (Harlan and Zohary 1966; Braidwood 1967). Braidwood later discovered other evidence of early agricultural practices at his excavations at Çayönü in the 1960s and 1970s. The site, located in southeast Turkey, successfully verified the extension of the nuclear zone into this region.

From the 1990s, an alternative view on the origin of agriculture came to the fore. This new hypothesis proposed that the first cultivation of cereals and legumes began in the Levant in the last phase of the Pleistocene. This period is archaeologically labeled as the Late Natufian period in the 11th millennium cal. BC and is represented, for example, by Natufian occupations at Abu Hureyra in Syria and Hayonim Cave and Ain Mallaha in Palestine (Bar-Yosef and Belfer-Cohen 2002; Hillman et al. 2011). This hypothesis argued that the first attempt of plant domestication was triggered by a global climate deterioration called the Younger Dryas event, which severely affected local plant habitats. It was argued that this rapid cooling and drying episode caused a recession of woodlands and a shortage of edible large-seeded annual plants, including cereals and legumes, which were already an important food resource for hunter-gatherer populations living in this area. To compensate for this food shortage, people began to cultivate wild plants, such as rye, wheat, barley and lentils, which had previously only been gathered from wild stands. This change caused a more intensive exploitation of cereals and legumes that eventually led to their morphological domestication. Supported by the paleo-climate study of the Younger Dryas, which is one of the most well-known examples of abrupt climate change in human history, this hypothesis was welcomed by many researchers, particularly at a time when ecological evolutionary thought was favored in archaeological studies.

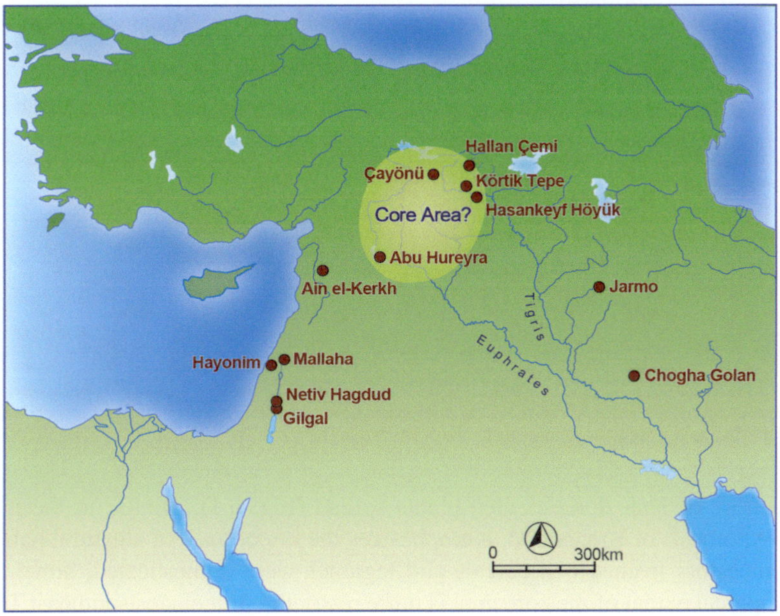


**Fig. 6.1** Map showing the geographical locations of the core area and the sites mentioned in the text

However, in the following decades, this hypothesis failed to provide firm evidence to support its argument. The increasing body of archaeobotanical evidence has instead suggested that there is no signs of an increasing use of cereals and legumes during the Late Natufian period and that their increase instead occurred as the climate upturned after the Younger Dryas (Willcox et al. 2009). Another study suggested that the Late Natufian society was marked by a return to the previous hunter-gatherer way of life, where a broader range of plant resources was utilized as a major subsistence strategy, rather than a more intensive use of cereals and legumes (Colledge and Conolly 2010; Rosen and Rivera-Collazoa 2012). Furthermore, refinement of dating techniques has recently shown that the Younger Dryas event and the social change in the Late Natufian period did not actually coincide in the first place (Blockley and Pinhasi 2011).

## 6.2 The Core Area Hypothesis

In the late 1990s, another hypothesis, which highlighted north Syria and southeast Turkey as the origin of plant domestication, was proposed based on the modern distribution of wild plant habitats and the results of a DNA study of modern einkorn wheat. This explanatory model is known as the "core area hypothesis" (Lev-Yadun et al. 2000; Abbo et al. 2010) and argues that the first domestication event occurred around the 9th millennium cal. BC within a "core area" where the habitation of wild

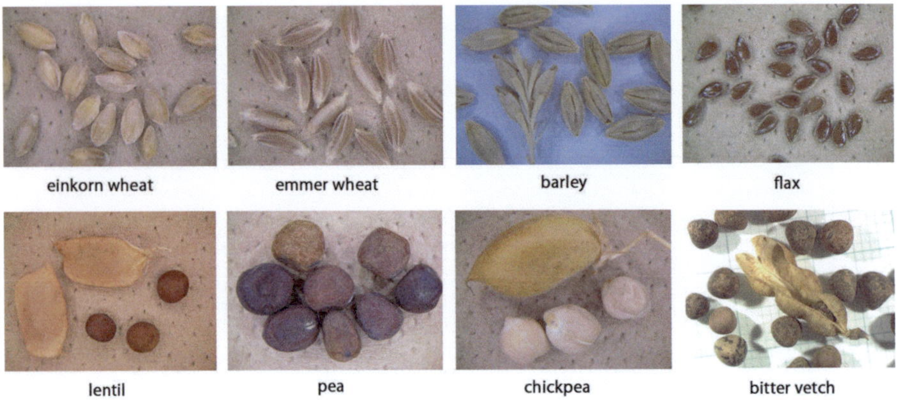


**Fig. 6.2** Seven founder crops and flax

progenitors of later domesticated plants spread (Fig. 6.1). Similar to the nuclear zone hypothesis of Braidwood, it emphasizes the importance of a natural habitat of wild ancestors from which cereals and legumes were domesticated, however the geographical extent of the core area is more restricted. It includes southeast Turkey, where Braidwood conducted his investigation in his later career, but does not include the Zagros foothills. In addition, the core area hypothesis emphasizes the role of human groups who are supposed to have started cultivating seven primary crops at once, rapidly developing a subsistence economy based on agriculture.

The seven crops, often called founder crops, are einkorn wheat, emmer wheat, barley, lentil, chickpea, pea and bitter vetch (Fig. 6.2). Flax is also included in the original definition of the eight founder crops by Zohary (1996), but recent studies show that the evidence of flax is rare and appears later in the archaeobotanical record. When looking at the modern distributions of wild populations of these crops, they spread to the Fertile Crescent and Anatolian plateau but to a different extent. All overlaps occur only in the area extending from the middle-upper Euphrates in north Syria to the upper Tigris in southeast Turkey. In particular, the natural habitat of wild chickpea is today only found in a limited area in southeast Turkey. This makes it plausible to suggest that the core area is the most possible candidate for the intensive exploitation of these crops that resulted in their domestication.

Furthermore, a DNA study (amplified fragment length polymorphism "finger-printing" analysis with the neighbor-joining phylogenetic tree method) of modern einkorn wheat has identified its origin in the Karacadağ mountain area in southeast Turkey (Heun et al. 1997). Here the analysis was conducted by calculating genetic distances among modern domestic and wild einkorns collected in West Asia and Europe. Phylogenetic tree diagrams (branching diagrams that show the evolutionary inter-relationships among analyzed samples) constructed in this analysis indicate that all domestic einkorns are genetically closely related, and a joined branch of domestic einkorns is joined with wild einkorn lines from the Karacadağ region. This means that modern einkorns sampled in West Asia and Europe originate from a single wild progenitor in the Karacadağ area. This result has been taken by the core area hypothesis as the evidence that underpins the argument for a single origin of agriculture.

The core area hypothesis was widely accepted by archaeologists particularly when other archaeological material evidence, such as architectural buildings and chipped stone tools, shows that many cultural elements in this period (the late Pre-Pottery Neolithic A and the early Pre-Pottery Neolithic B) also diffused from the core area to other regions of West Asia.

## 6.3 Critique of the Single Origin of Crop Domestication

However, evidence from recent archaeobotanical studies has identified serious flaws in the core area hypothesis. The core area hypothesis has been criticized because it is exclusively based on modern botanical collections and assumes that modern plant distributions and their DNA traits are representative of those of the past. In reality, the past habitat of founder crops may have been different from present ones; in fact, some archaeobotanical evidence shows that the earliest examples of founder crops were recovered from outside the core area.

In addition, the computer simulation used to perform the phylogenetic analysis described above has been severely criticized (e.g., Allaby et al. 2008). It has been argued that the evolutionary statistics used in the DNA study of einkorns are unreliable because simulated data frequently infer a single origin, even if more than one origin were input as primary datasets for the simulation. This was because it was incorrectly simulated by the algorithm that did not consider hybridization events. This undermines the scientific basis on which the core area hypothesis relies.

Furthermore, it has long been known from a cross-allelism test of hybrids that the most important morphological trait of domestic-type barley, that is, non-shattering rachis (see below), is controlled by two genes, *btr1* and *btr2*, which have different geographical distributions (Takahashi 1955). This basic genetic study provides crucial evidence for the multiple origins of domestic-type barley.

An alternative model for the origin of agriculture has been proposed by primarily relying on archaeobotanical evidence (Tanno and Willcox 2006a; Fuller 2007). Based on recently enriched archaeobotanical remains this new hypothesis argues that the domestication of crops was a protracted process that developed in more than one place in West Asia, where cultivation of wild local plants was attempted through trial and error over several thousand years.

## 6.4 Protracted Process of Crop Domestication

One of the key elements of this hypothesis is to clarify when and where morphologically domestic-type crops appeared based on the secure identification of wild and domestic crops among archaeobotanical remains recovered from excavations. To

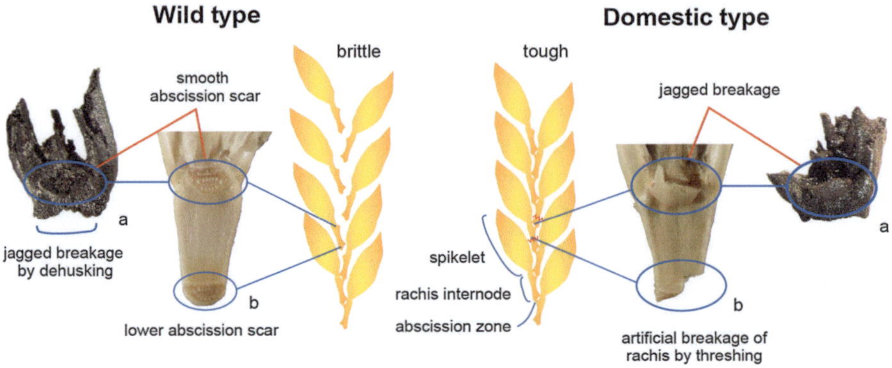


**Fig. 6.3** Morphological differences of wild-type and domestic-type rachises of hulled wheat. (**a**) Charred archeobotanical samples and (**b**) modern samples

distinguish whether charred remains of wheat and barley are wild or domestic, morphological features of their rachises were examined (Tanno and Willcox 2006a, 2012).

The rachis is the main axis on the ear of cereals, to which rows of spikelets (within which seed grains grow) are attached (Fig. 6.3). Wild-type cereals have brittle rachises, and when they have fully ripened, the rachis is naturally disarticulated; thus, the spikelets, to which a short segment of the rachis (rachis internode) is attached are shattered from the ear. This is the mechanism of wild cereal seed dispersal. As a result of natural shattering, a rachis internode attached to a spikelet has a smooth abscission scar at the location where rachis disarticulation occurred (Fig. 6.3). On the other hand, domestic-type cereals have a tough rachis caused by their mutation. A tough rachis does not disarticulate at maturity, and spikelets are not shattered from the ear. A non-shattering ear is easier to harvest because it prevents the loss of grains before and during harvesting. Separation of the spikelets is achieved by threshing after harvesting, and this artificial separation leaves jagged breakage on domestic-type rachises instead of smooth abscission scars.

Therefore, this morphological difference between wild-type and domestic-type rachises, that is the presence of smooth or jagged abscission scars, is the most reliable criteria to distinguish between wild and domestic varieties of wheat and barley. However, the identification requires careful examination of rachises because they are often fragmented and a jagged breakage may occur during the de-husking process. This could occur when de-husking wild-type spikelets, and they must be differentiated from genuine domestic-type rachises.

Careful examination of rachis types from samples recovered from several early Neolithic sites occupied from the 9th millennium cal. BC onward were analyzed. As a result, the chronological change in the relative frequencies of wild and domestic hulled wheat (i.e., einkorn and emmer wheat) and barley has become apparent and demonstrated that the domestic type increased very slowly through time, taking more than a few thousand years to dominate. Figure 6.4 shows that the percentage of domestic hulled wheat began to increase after 8500 cal. BC but did not outnumber the wild-type hulled wheat until more than a millennium later. Similarly, domestic-

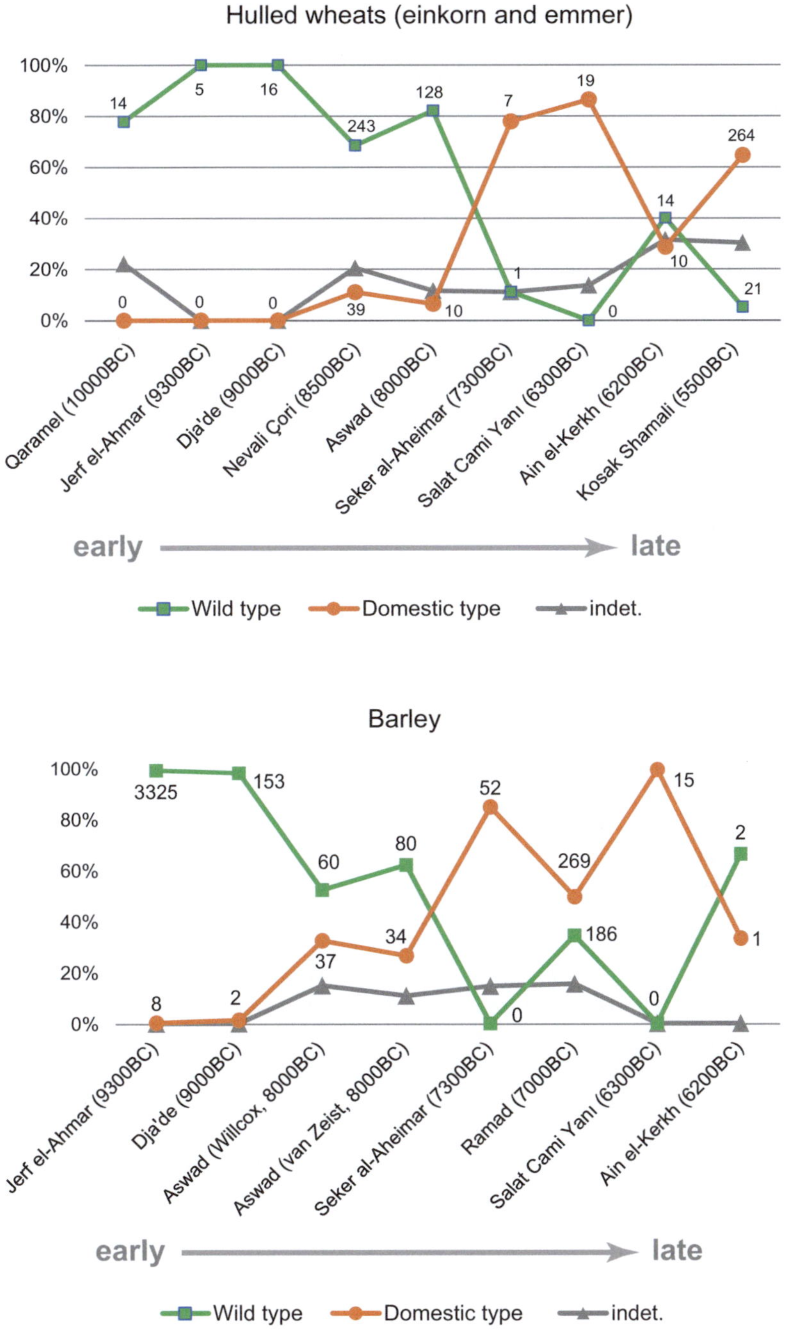


**Fig. 6.4** Chronological change in the relative frequencies of wild-type and domestic-type rachises. *Top*: hulled wheat (i.e., einkorn and emmer) and *bottom*: barley. The numbers in the graph indicate the count of identified samples. "Indet." represents those with abscission scars that cannot be easily morphologically distinguished because of post-depositional damage. They do not have smooth abscission scars and were classified as "possible domestic" in Tanno and Willcox (2012)

type barley appeared around 9000 cal. BC but did not become dominant until about 7300 cal. BC (Fig. 6.4). This protracted process means that the wild type was used along with the domestic type for an extended period, and the gradual increase in the domestic type was the unintended consequence of human cultivation practices. This contradicts the rapid domestication model proposed by Hillman and Davis (1990), who argue that morphologically domestic types become dominant within 20–200 years under strong pressure of human selection. The core area hypothesis, which also believes that domestication of crops was driven by a conscious selection of domestic types by the people living in the core area (Abbo et al. 2010), must be equally questioned.

## 6.5 Multiple Origins of Crop Domestication

The slow increase of domestic cereals at the same time suggests that the first crop cultivation may have already occurred a few thousand years before the appearance of morphologically domestic rachises, which appear in the mid-9th millennium cal. BC (Tanno and Willcox 2006a, 2012). The cultivation of "morphologically wild" cereals (often referred to as pre-domestication cultivation) is difficult to identify in archaeological records but is a crucial issue for the origin of agriculture. For example, possible evidence of pre-domestication cultivation is known from the south Levant, far outside the core area. At Gilgal, which was occupied during the Pre-Pottery Neolithic A period in the late 10th to early 9th millennia cal. BC, a substantial quantity of wild barley and oat (more than 260,000 wild barley grains combined with 120,000 wild oats) was stored in a storage facility (Weiss et al. 2006). Its quantity far exceeds the yield that would have been expected from harvesting a local natural population, and the practice of pre-domestication cultivation is expected. A similar stash of barley seeds is also known from the adjacent contemporary site of Netiv Hagdud. The possibility of wild cereal cultivation has also been proposed at several sites in the middle Euphrates in north Syria (within the core area), where the exploitation of wild cereals increased before domestication (Willcox et al. 2008). Furthermore, the recent discovery of possible cultivated wild barley at Chogha Golan in the Iranian Zagros also provides evidence for the practice of pre-domestication cultivation outside the core area (Riehl et al. 2013). In an earlier level of Chogha Golan, dating back to the early 9th millennium cal. BC, the proportion of wild barley increases together with arable weeds that usually grow in a cultivated field. Domestic-type cereal appears in a later level of this site and implies a local process of cereal domestication in an area more than 1000 km away from the core area (Fig. 6.1).

Multiple origins of crop domestication are also supported archaeobotanically by the fact that more than seven founder crops were cultivated at the beginning of the Neolithic. First, Fuller et al. (2011) pointed out that a wide range of crop species and subspecies that no longer exist today were cultivated at the onset of agriculture. For example, two-grained einkorn has been recovered from several early Neolithic sites

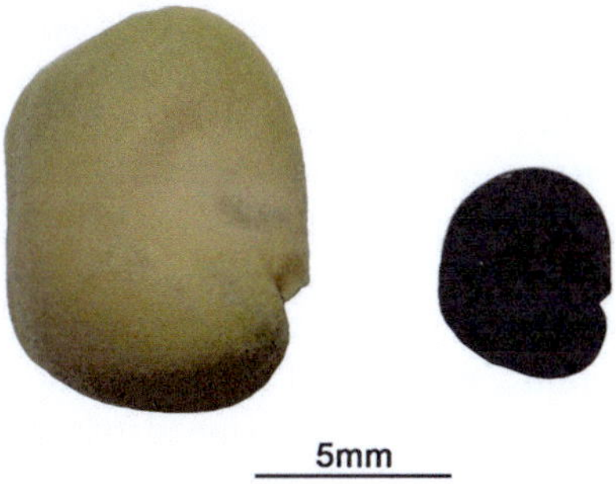


**Fig. 6.5** Faba beans. *Left*: a modern domestic seed. *Right*: the earliest definitive evidence of charred faba bean from the Early Pre-Pottery Neolithic B levels at Tell Ain el-Kerkh (ca. 8500 cal. BC). It is flat and somewhat thick at the hilum's side

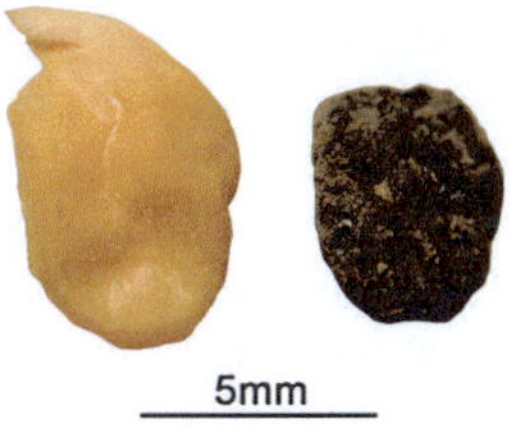


**Fig. 6.6** Chickpeas. *Left*: a modern wild seed. *Right*: a charred ancient seed with wild-type morphology from the Early Pre-Pottery Neolithic B levels at Tell Ain el-Kerkh (ca. 8500 cal. BC). Wild-type morphology is only known from Tell Ain el-Kerkh, where both wild and domestic types have been identified

in the south Levant as well as in the core area. Striate emmeroid wheat (new type glume wheat) has been recovered from early sites in Anatolia and Greece. These species are not included in the archaeobotanical remains in the later age, and no modern domestic population is known. Second, the evidence of faba beans recovered from the Early Pre-Pottery Neolithic B level (ca. 8500 cal. BC) of Tell Ain el-Kerkh in northwest Syria (Fig. 6.5) shows that faba bean, which is not included in the founder crops, has also been an important food crop since the late 9th millennium cal. BC (Tanno and Willcox 2006b). The exploitation of a great amount of oat at Gilgal mentioned above also indicates that the founder crops were not exclusive food sources for Neolithic people. The focus on only studying the seven founder crops must be criticized as an insufficient strategy to fully identify the original locales of plant domestication.

Furthermore, the main argument in support of the core area hypothesis, claiming that the natural habitat of wild chickpea progenitors was limited to the core area, must be questioned. This is because the oldest evidence of wild- and domestic-type chickpeas (Fig. 6.6) has so far been recovered, together with faba beans, from the Early Pre-Pottery Neolithic B level at Tell Ain el-Kerkh, which is far from the core area (Tanno and Willcox 2006b). The modern habitat distributions of plants may have changed since the onset of agriculture and could be misleading if ignoring the data recovered from archaeological excavations.

While all these archaeobotanical data provide examples that indicate the emergence of crop domestication outside the core area, new archaeological investigations of early Neolithic sites within the core area itself demonstrate a lack of evidence of crop domestication at some sites. In the eastern part of the upper Tigris valley in southeast Turkey, excavations currently being conducted at Early Neolithic (Pre-Pottery Neolithic A) sites have revealed that these settlements were sedentary villages of hunter-gatherers who did not practice crop cultivation. Among them is Hasankeyf Höyük that has been investigated by the University of Tsukuba since 2011 (Miyake et al. 2012). The site dates back to the second half of the 9th millennium cal. BC, the period during which pre-domestication cultivation or initial domestication was assumed to be practiced at many sites in West Asia. Nevertheless, archaeobotanical remains from Hasankeyf Höyük show no signs of cereal cultivation. Pistachio, almonds and hackberry are the major plant taxa at this site, and wheat and barley, whether wild or domestic, are virtually absent. Similarly, the very low frequency of cereals is observed at the neighboring contemporary site of Körtik Tepe (Riehl et al. 2012). Also, the recent analysis of plant remains from the Hallan Çemi equally shows that cereals were rarely used at this site (Savard et al. 2006). Interestingly, no occupational evidence from the succeeding Pre-Pottery Neolithic B period has so far been attested in this area, and evidence of full domestication of crops appears in the Pottery Neolithic period when settlements were re-introduced to the area. Located only about 100 km east of Mount Karacadağ, the upper Tigris valley is included in the core area. However, it seems likely that this region did not experience the local development of plant domestication.

## 6.6 Conclusions

Studies based on the latest archaeobotanical evidence suggest that the development of crop domestication was a prolonged process occurring over a few millennia at multiple locales. The first agricultural practice, that is, the cultivation of wild-type species or pre-domestication cultivation, is believed to have begun at multiple locations over a wide area of West Asia, including the south Levant and the Zagros foothills in addition to the core area, at least by the late 10th to early 9th millennia cal. BC. The succeeding development of agriculture accompanied by the emergence and increase in morphologically domestic-type crops gradually occurred at various places following different trajectories according to the natural and cultural circumstances in each region.

In fact, the core area hypothesis is correct in emphasizing the genealogical significance of the core area. The core area is certainly one of the most important areas for the agricultural origin, and the archaeobotanical evidence ratifies its importance rather than denying it. However, the critique of the core area hypothesis pertains to its insistent proclamation of the "exclusive role" of the core area as the single origin of agriculture, which does not fit the latest archaeobotanical evidence.

It seems more likely that the dawn of agriculture was neither an invention made by a particular group of people nor an event that occurred at a particular time and place in response to economic or social necessities. Rather, it is more plausible to regard it as a process of transformation that was developed through many generations in terms of both the genetic evolution of plants and accumulation of knowledge and experience by human beings. In due course, this transformation reached the "point of no return" and became the substantial economic foundation on which modern civilization was later established.

## References

Abbo, S., S. Lev-Yadun, and A. Gopher. 2010. Agricultural origins: Centres and non-centres: A Near Eastern reappraisal. *Critical Reviews in Plant Sciences* 29: 317–328.

Allaby, R.G., D. Fuller, and T.A. Brown. 2008. The genetic expectations of a protracted model for the origins of domesticated crops. *Proceedings of the National Academy of Sciences* 105(37): 13982–13986.

Bar-Yosef, O., and A. Belfer-Cohen. 2002. Facing environmental crisis. Societal and cultural changes at the transition from the Younger Dryas to the Holocene in the Levant. In *The dawn of farming in the near east*, Studies in early near eastern production, subsistence, and environment 6, ed. R.T.J. Cappers and S. Bottema, 55–66. Berlin: Ex oriente.

Blockley, S.P.E., and R. Pinhasi. 2011. A revised chronology for the adoption of agriculture in the southern Levant and the role of Lateglacial climatic change. *Quaternary Science Reviews* 30: 98–108.

Braidwood, R.J. 1967. *Prehistoric men*, 7th ed. Glenview: Scott Foresman.

Braidwood, R.J., and B. Howe. 1960. *Prehistoric investigations in Iraqi Kurdistan*, Studies in ancient oriental civilization, No.31. Chicago: The Oriental Institute of the University of Chicago.

Braidwood, L.S., R.J. Braidwood, B. Howe, C.A. Reed, and P.J. Watson. 1983. *Prehistoric archaeology along the Zagros flanks*, Oriental Institute Publications 105. Chicago: The Oriental Institute of University of Chicago.

Colledge, S., and J. Conolly. 2010. Reassessing the evidence for the cultivation of wild crops during the Younger Dryas at Tell Abu Hureyra, Syria. *Environmental Archaeology* 15: 124–138.

Fuller, D. 2007. Contrasting patterns in crop domestication and domestication rates: Recent archaeobotanical insights from the Old World. *Annals of Botany* 100: 903–924.

Fuller, D.Q., G. Willcox, and R.G. Allaby. 2011. Cultivation and domestication had multiple origins: Arguments against the core area hypothesis for the origins of agriculture in the near east. *World Archaeology* 43: 628–652.

Harlan, J.R., and D. Zohary. 1966. Distribution of wild wheat and barley. *Science* 153: 1074–1080.

Heun, M., R. Schäfer-Pregl, D. Klawan, R. Castagna, M. Accerbi, B. Borghi, and F. Salamini. 1997. Site of einkorn wheat domestication identified by DNA fingerprinting. *Science* 278: 1312–1314.

Hillman, G.C., and M.S. Davies. 1990. Domestication rates in wild-type wheats and barley under primitive cultivation. *Biological Journal of the Linnean Society* 39: 39–78.

Hillman, G., R. Hedges, A. Moore, S. Colledge, and P. Pettitt. 2011. New evidence of Lateglacial cereal cultivation at Abu Hureyra on the Euphrates. *The Holocene* 11(4): 383–393.

Lev-Yadun, S., A. Gopher, and S. Abbo. 2000. The cradle of agriculture. *Science* 288: 1602–1603.

Miyake, Y., O. Maeda, K. Tanno, H. Hongo, and C.Y. Gündem. 2012. New excavations at Hasankeyf Höyük: A 10th millennium cal. BC site on the upper Tigris, southeast Anatolia. *Neo-Lithics* 1(12): 3–7.

Riehl, S., M. Benz, N.J. Conard, H. Darabi, K. Deckers, H.F. Nashli, and M. Zeidi-Kulehparcheh. 2012. Plant use in three Pre-pottery neolithic sites of the northern and eastern fertile crescent: A preliminary report. *Vegetation History and Archaeobotany* 21: 95–106.

Riehl, S., M. Zeidi, and N.J. Conard. 2013. Emergence of agriculture in the foothills of the Zagros mountains of Iran. *Science* 341: 65–67.

Rosen, A.M., and I. Rivera-Collazoa. 2012. Climate change, adaptive cycles, and the persistence of foraging economies during the late Pleistocene/Holocene transition in the Levant. *Proceedings of the National Academy of Sciences* 109(10): 3640–3645.

Savard, M., M. Nesbitt, and M.K. Jones. 2006. The role of wild grasses in subsistence and sedentism: New evidence from the northern fertile crescent. *World Archaeology* 38(2): 179–196.

Takahashi, R. 1955. The origin and evolution of cultivated barley. In *Advances in genetics Vol.7*, ed. M. Demerec, 227–266. New York: Academic.

Tanno, K., and G. Willcox. 2006a. How fast was wild wheat domesticated? *Science* 311: 1886.

Tanno, K., and G. Willcox. 2006b. The origins of cultivation of *Cicer arietinum* L. and *Vicia faba* L.: Early finds from northwest Syria (Tell el-Kerkh, late 10th millennium BP). *Vegetation History and Archaeobotany* 15: 197–204.

Tanno, K., and G. Willcox. 2012. Distinguishing wild and domestic wheat and barley spikelets from early Holocene sites in the near east. *Vegetation History and Archaeobotany* 27: 107–115.

Weiss, E., M.E. Kislev, and A. Hartmann. 2006. Autonomous cultivation before domestication. *Science* 312: 1608–1610.

Willcox, G., S. Fornite, and L. Herveux. 2008. Early Holocene cultivation before domestication in northern Syria. *Vegetation History and Archaeobotany* 17: 313–325.

Willcox, G., R. Buxo, and L. Herveux. 2009. Late Pleistocene and early Holocene climate and the beginnings of cultivation in northern Syria. *The Holocene* 19(1): 151–158.

Zohary, D. 1996. The mode of domestication of the founder crops of southwest Asian agriculture. In *The origins and spread of agriculture and pastoralism in Eurasia*, ed. D. Harris, 142–158. London: UCL Press.

# Chapter 7
# The "Commodification" of Animals

**Hans-Peter Uerpmann and Margarethe Uerpmann**

## 7.1 Introduction

The basic precondition for the transformation of animals into "commodities" – defined as useful articles or facilities – is their immediate availability for the intended purpose. This is certainly not the case for animals living free in the wild. Domestication – or at least taming – is therefore the basic precondition for human use of live animals for whatever purpose. Research into the origins of animal domestication has a long tradition, which began with Charles Darwin (1868) and continues to the present day (e.g. Clutton-Brook (1999); Hongo et al. (2009); Conolly et al. (2011)). Definitions of domestication vary, but there is basic agreement that – in biological terms – domestication can be understood as separation of a domestic sub-population from the original wild population of the same species. Domestication must therefore be seen both as an anthropogenic and a biological process.

The enormous increase in phenotypic variation of domestic animals in comparison to their wild ancestors is the most evident difference between the wild forms and their domestic offspring. Biological alterations, in particular changes in skeletal morphology as well as in their genetic signatures, enable the archaeo-zoologist and palaeo-geneticist to discriminate between the remains of domestic and wild animals. It has become common knowledge that Western Asia was a major center of animal domestication in prehistoric times. Nevertheless, the basic questions of why, when and where exactly this process began, are still under debate. Often a utilitarian approach is applied, assuming that animals were domesticated for particular purposes – like providing meat, milk, skin, wool, labor, or other commodities. If this really had been the case the general question arises why animal domestication did

H.-P. Uerpmann (✉) • M. Uerpmann
Institute of Scientific Archaeology, Tübingen University,
Rümelinstrasse 23, Tübingen 7207, Germany
e-mail: hans-peter.uerpmann@uni-tuebingen.de



A. Tsuneki et al. (eds.), *Ancient West Asian Civilization*,
DOI 10.1007/978-981-10-0554-1_7

not happen much earlier during the development of Early Modern Humans. In particular, consideration should be given to why the early inhabitants of West Asia were the first to develop such a simple idea for making human life so much easier. Last but not least it is also necessary to determine the conditions under which the dog was the first animal to be domesticated. It eats the same food as humans and is usually not eaten by its human companions, which means that it costs more than it brings in.

Actually, animal domestication was not a simple invention at all, and it was not identical from one domestic animal species to the other. Rather, it was a gradual process which began more than 15,000 years ago and requires a number of preconditions on the human, animal, and environmental side. The result of domestication is an inclusion of particular animals into the human sphere.

## 7.2 The Beginnings of Animal Domestication

One of the oldest presently known finds of an unequivocally domestic animal is the fragment of a dog-skull (Fig. 7.1) from Kesslerloch Cave in northern Switzerland, where this dog lived more than 14,000 years ago (Napierala and Uerpmann 2010). Domestication of the wolf can be imagined as a comparatively simple process which could have happened independently in different parts of the world. Genetic evidence, however, indicates that all domestic dogs worldwide originated in Europe some 15,000–20,000 years ago (Thalmann et al. 2013). A Late Paleolithic dog was also identified at Kebarah cave (Dayan 1994) in Israel.

The most likely scenario of wolf domestication begins as an unconscious process. Occasional fostering of an orphaned wolf puppy in a human community leads to an emotional inclusion of the animal into the human sphere and to an acceptance of humans as "conspecific" on the animal side. Such man-animal relationships are usually fragile and may easily break up when the animals raised in a human community reach sexual maturity and look for a sexual partner. However, in the case of raised young wolves both sexes may sometimes have lived simultaneously within a group of Paleolithic hunters. Their potential offspring would – from the beginning – have grown up in human company. This would have made them real members of the respective groups. Under such conditions and isolated from wild wolves they will soon have developed diverging pheno- and genotypes, and would have become real ***dogs***. This would have happened without conscious interaction from the human members of the respective community. Thus, self-domestication is a realistic option for the beginnings of the close relationship between humans and dogs. The "commodities" provided to a group of hunter-gatherers by the possession of dogs – e.g. their superior sense of smelling – were manifold, and certainly sufficient to keep this relationship going even beyond the era of the hunter-gatherers. It is essential to realize that this early case of animal domestication was not an "invention." It rather was a natural development which, because of its extension over several generations

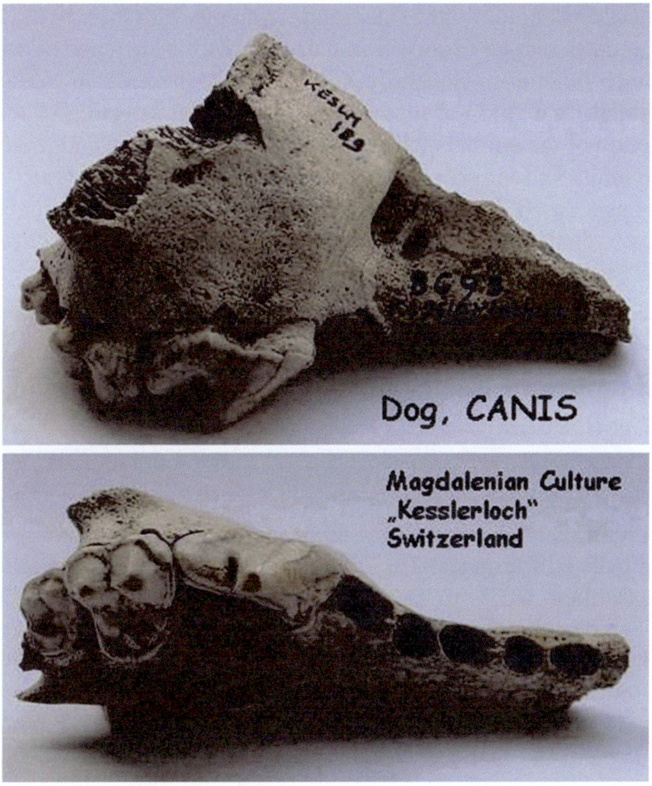


**Fig. 7.1** Paleolithic dog-skull from Kesslerloch cave

of “wolf/dogs,” would not have been realized by the human participants as a sort of revolution. Therefore the domestication of the dog would not have been registered as a process that also could be transferred to other animal species. However, the possession of dogs was obviously considered an advantage by the Late Paleolithic hunters, leading to a general spread of dogs among the Late Pleistocene hunter-gatherers of northern Eurasia from where it was also taken to America.

At this point it is necessary to deal with the biological meaning of domestication. Seen from a biological perspective, domestication circumscribes a range of processes, which leads to the separation of a *domestic* sub-population from the original *wild* population of the same species. Therefore domestication must not only be seen as a human achievement but also as a *biological process*, which was *induced by humans*. Generally speaking the basic biology of domestication has to be understood before human participation in this process can be dealt with.

The major biological effect of domestication is an enormous increase of phenotypic variation of the animals under concern. Wolves for example look more or less the same in their huge range throughout the northern hemisphere from the Iberian Peninsula in the west to Mexico on the other side of the world. Dogs on the other hand have all sorts of colors and a size- range from tiny Pinschers to the Great Dane. Similar differences can be observed in cats, but also in the species which mainly have economic functions. Generally speaking this increase in variation among domestic animals is due to the replacement of "natural selection of the fittest" by human selection of the animals which were suitable for living and reproducing in a human community and for providing certain products or commodities. The biological differences caused by the diverging criteria of selection enable the bio-archaeologist to distinguish the remains of wild and domestic animals provided that these differences are reflected in the skeleton.

## 7.3 Economic and Ecological Aspects of Early Animal Domestication

The basic question for bio-archaeology and human history in general of why animals were domesticated after many thousand years of human history without domestic animals still remains. The obvious answer seems to be because humans wanted to ease their access to meat and other animal "commodities" (e.g. Davis 2005). However, this is not plausible. If this had been the reason, the Late Paleolithic hunters in Europe and North Asia should have domesticated the reindeer. It was their major source of meat and it had the behavioural capacity to become domesticated. Humans already had the dog as an example of a domestic animal. Further questions surround the fact that the dog, the first domesticate, was not regularly eaten. It is strange that further domestications did not follow immediately, once the basic principles of domestication were known. The answer can only be that the potential of animal domestication as a possibility to get access to meat and other animal products was beyond the imagination of hunters and gatherers. It seems likely that Paleolithic people, even though they had dogs, did not realize the economic potential behind this form of man-animal relationship. Utilitarian and retrospective approaches to the domestication processes should therefore be seen with great caution when applied to the initial processes of early plant cultivation and animal domestication. A Paleolithic hunter could not have imagined that his animals of prey would not run away when they saw him coming.

Economy and ecology of the Late Pleistocene hunters and gatherers have to be considered in detail in order to understand their ecological situation at the time when plant cultivation and the domestication of animals with an economic potential began some 12,000 years ago. It is interesting in this context that early cultivation and domestication processes were not restricted to Southwest Asia. Actually, there were three centers of early plant cultivation located in West Asia, East Asia, and Central America. They were separated from each other by huge distances – mountains, deserts and oceans. In Central America the cultivation of maize started at roughly the same time as cereal cultivation began in West Asia. Indications for the contemporary beginnings of rice cultivation and pig domestication have been reported from China. It is impossible that these coeval developments were initiated through cultural exchange between the respective human populations. The only possible explanation for the contemporaneity of these early cultivation and domestication processes are the global climate changes at the transition from the last Ice-Age to the Holocene. However, the big question of how – and why – climate change could trigger processes like plant cultivation and animal domestication in some parts of the world and not in others still remains.

An in-depth discussion of the problems related to this question is not possible here. It is an obvious fact, however, that climate change, which is never a fast event, does not directly affect individual humans, animals, or plants. It rather affects the dynamics of whole populations. Global warming at the end of the Ice-Age meant more evaporation over the oceans resulting in more precipitation inland. This was when the so-called Fertile Crescent of West Asia actually became fertile. Similar processes must have happened in the other two centers of early domestication. The tall and large-seeded grasses, wild wheat, barley and rye in South-West Asia, rice in East Asia and maize in Central America will have benefitted from this climate change. Their proliferation with increasing precipitation was faster than that of the trees and bushes, which otherwise suppress the grasses by casting their shade on them. For a considerable time after the onset of higher temperatures and increased precipitation the Fertile Crescent, being a treeless steppe during the peak of the glacials, must have been flourishing grasslands – a paradise for all those beings that could live on grasses and cereals. Similar processes can be imagined for the areas where wild rice or maize existed.

The availability of more cereal-grains than a local group of hunters and gatherers could eat during the period when they were ripe is the next link in the chain of processes leading to the evolution of a new scheme of human subsistence. Storing the surplus of grains was a self- evident reaction, which had further consequences for human life-style. The stocks of grain necessary for a whole year would have been too heavy to carry for a group of mobile hunters and gatherers. Possibilities for storage will soon have been discovered. But nevertheless these stocks restricted mobility, because they had to be protected and were too heavy to transport. Partial sedentism was required for this purpose, a clear disadvantage for hunters who had to follow their prey. On the other hand, sedentism was facilitated by the availability of stored food. All in all this is a model of how the development of "sedentary hunters and gatherers" might have taken place. The Natufian Culture of the Levant is paradigmatic for the first evolutionary step towards "developed societies" in West Asia.

This process was continued by the development of agriculture which can easily be imagined as a consequence of the first step. The storage of cereal grains in pits in the ground will soon have led to the observation that a new plant will grow out of a lost grain. Seeding and the gradual development of agricultural techniques would be self-evident consequences which lead to a more sedentary life-style. The development of complex architecture will have been the next step, well documented in the West Asian Levant by remnants of large settlements of the Late Natufien and the Pre-Pottery Neolithic A (PPN-A).

The domestication of meat-producing animals was the next step in the evolution of human economy. The close connection of incipient agriculture and the domestication of sheep and goats is documented by the fact that evidence for the latter process is only found in restricted parts of the original distribution areas of these animals (Figs. 7.2 and 7.3; see also Uerpmann 1987).

Early animal domestication is documented both by changing frequencies of bone remains of domestic sheep and goats at the settlements of the Pre-Pottery Neolithic (PPN-B) and by morphological changes of their skeletons. However, the large Levantine sites do not provide much evidence for the domestication process itself. This may be due to the fact that the actual domestication process happened before the economic developments took place which led to an increase of the human population.

Sheep and goats were the earliest meat-producing animals to become domesticated. How this actually happened is a matter of conjecture rather than of archaeobiological research. The biological processes necessary to produce distinctive morphological features in the skeletons of early domesticates would have become manifest within less than 50 generations, a time- span of no more than about 200 years. This is generally below the temporal resolution of excavation techniques, even at well stratified Neolithic Tell-sites in Western Asia. The scenario to be imagined is an unconscious process of fostering and taming of young lambs and kids of wild sheep and goats, found in the wild or orphaned by hunting their mothers.

Obviously this imaginative scheme of ungulate-domestication gives raise to the question why this process could not have happened much earlier. The answer is that lambs or kids – unlike wolf-puppies – would not have been able to follow a group of mobile hunters and gatherers. In order to find enough to eat, they needed a lot of time while moving around in the landscape. Thus, a high degree of sedentism, the existence of a *domus,* is a precondition for the *dom*estication of plant-eating animals. This does not exclude the later development of mobile herding societies. Their speed of translocation would already have been adapted to the biology of the herded animals.

The chain of successive developments can thus be lined out as (1) climate change towards more precipitation, (2) fast proliferation of wild cereals in the still tree-less steppes, (3) collection of a surplus of cereal grains and invention of grain storage, (4) increasing sedentism in order to protect the stored provisions combined with continued hunting of wild game for protein. (5) Kids and lambs of wild sheep and goats, orphaned through the killing of their mothers during the hunt and occasionally brought back to the incipient settlements, would have become tame and some of them will have survived till sexual maturity. This would not have been possible in a community of mobile hunters and gatherers, because – other than a wolf-puppy – young lambs or goat-kids would not have been able to follow a wandering human group, because they would have had to stop frequently for browsing. As a general conclusion it can be stated that (6) the domestication of plant-eating animals required a sedentary way of life of the domesticators. Once food-production, both on the vegetal side for the production of carbohydrates for human nutrition, and on the animal side with regard to protein provision, a base for the establishment of self-controlled subsistence systems was laid.

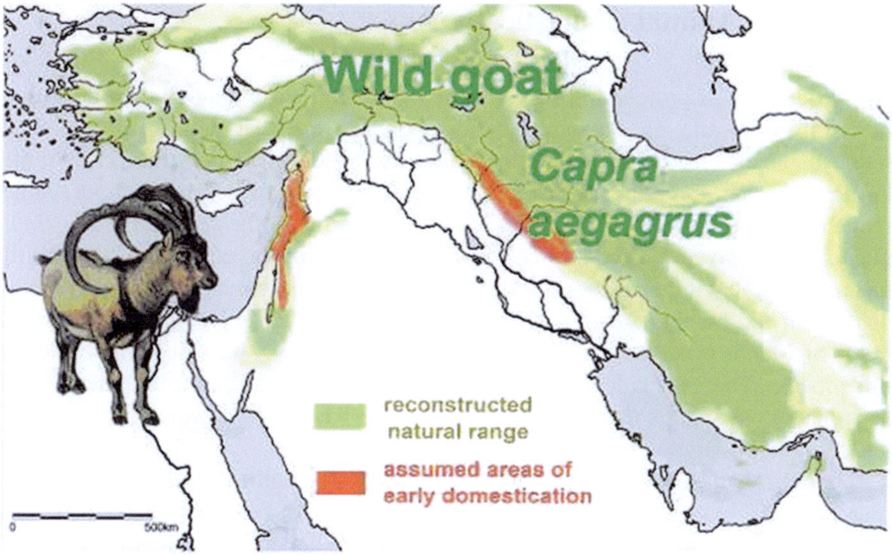


**Fig. 7.2** Ancient distribution and areas of early goat domestication

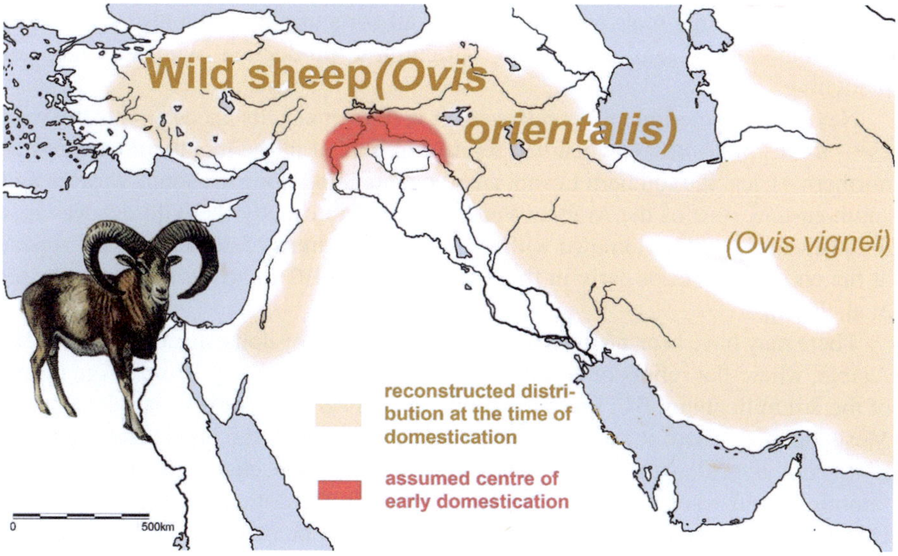


**Fig. 7.3** Ancient distribution and area of early sheep domestication

Developed sedentism and the beginnings of urbanisation were the natural consequences of the processes described above. Intellectual pervasion of the role of sheep and goats for human subsistence may have enabled the inhabitants of these first rural settlements to induce further domestications. Pigs, and in particular cattle may have been domesticated consciously, thus broadening the base of human subsistence.

## 7.4 Later Animal Domestications

The production of protein and fat for human nutrition has remained the only function of the domestic pig. Its domestication soon followed that of sheep and goats. It seems to have happened in the mountainous areas bordering the northern arch of the Fertile Crescent. In contrast to pigs the domestication of cattle opened up a whole range of new potential "commodities." The first taming of the Aurochs (*Bos primigenius*) must have required sturdy buildings because of the enormous strength of this animal, the wild ancestor of domestic cattle. It would not have been possible to confine an adolescent Auroch in a simple wooden shed. Early Neolithic settlements in the highlands of Southern Anatolia and Western Iran may have been the center of this process. There is evidence from mitochondrial DNA isolated from bones found at archaeological sites that all represented cattle had a very restricted number of ancestors (Bollongino et al. 2012).

The murals at Catal Höyük (e.g. Mellaart 1967) might indicate that religious reasons could have played a role in the domestication of this most impressive animal. Proving this assumption will, however, remain beyond the scientific scope of bio-archaeology. The other commodities provided by cattle, milk and labor, might originally only have been "side- products." Nevertheless, the use of cattle as draught animals meant an enormous amplification of the human subsistence economy. It did not only allow large scale transportation of all sorts of goods, but also – and even more important – an increase in agricultural production with the help of ploughing.

Next on the time-line of domestications is the donkey. Its ancestor was the wild ass – *Equus africanus* – which had its natural habitat in the mountainous areas of northern Africa, the southern Levant and the mountains along the south-western and south-eastern rims of the Arabian Peninsula. Where exactly the wild ass was first domesticated is still not known with certainty. Early finds of domestic asses appear at the end of the 4th or early in the 3rd millennium BC in NE-Africa (e.g. Rossel et al. 2008).

There may have been an independent center of donkey domestication in southern Arabia, where flat reliefs of equids with riders appear on Early Bronze Age tombs of the 3rd millennium BC. From the same time onward the donkey was also used in Mesopotamia, where it seems to have been crossed with the hemione in order to produce a hybrid with special characters. Whether the chariot depicted on the famous "Standard of Ur" (Fig. 7.4); from the Early Dynastic period (c. 2850–2350 BC) was drawn by donkeys or such hybrids cannot be determined. Evidence for hybridisation is deduced from equid skeletons found in prominent burials (Weber 2008).

All the domestic ungulates dealt with up to now – sheep, goat, cattle and donkey – had their origin within the cultural sphere of the ancient civilisations of South West Asia. The horse, however, with its enormous impact on economic and cultural developments, is an "outsider". The natural range of wild horses – and hence the area where it might have been domesticated – fluctuated widely during the Holocene.

**Fig. 7.4** Chariot drawn by four donkeys or donkey-onager hybrids

These fluctuations depended on climatic changes. Wild horses are inhabitants of grass-steppes. The core areas of this habitat extend from eastern central Europe through south central Asia till Mongolia and eastern Siberia. During cool and dry phases, particularly those occurring in the 4th millennium BC, the range of wild horses extended southwards. In western Europe they reached the Iberian Peninsula and in south-eastern Europe they occupied the plains along the lower Danube, the northern shore-lands of the Black Sea and Trans-Caucasia. From there they even penetrated the central highlands of Anatolia and probably also those of western Iran. At times they might even have reached the eastern shore-lands of the Mediterranean Sea. These southern extensions of the habitat of wild horses provoked claims of an early appearance of domestic horses during the final Neolithic period. Such claims, which cannot easily be dismissed, are based on the occurrence of horses in areas where they did not exist before.

Other claims of horse domestication are based on the occurrence of a particular form of tooth abrasion, thought to be caused by a bit, the mouth-piece of a bridle. Such forms of abrasion actually do occur at the first lower cheek-teeth of riding-horses. But they also occur naturally when the corresponding tooth in the upper jaw is slightly more protruding than its lower counterpart. It is therefore necessary to have both the first upper and lower cheek-teeth of the respective animal in order to claim that it had a bit in its mouth.

Clear evidence for an early occurrence of domestic horses in SW-Asia is their appearance as grave-gifts in noble tombs (Fig. 7.5) of the early 3rd millennium BC in Transcaucasia (Uerpmann and Uerpmann 2010), where at that time the wild horse had disappeared again. From Transcaucasia the domestic horse spread south, reaching Mesopotamia during the later part of the 3rd millennium BC. Probably it

**Fig. 7.5** Selected bones of a juvenile and an adult horse and of a sheep deposited at the base of the deep grave-shaft of a noble burial at Nerkin Naver (Central Armenia)

took some time till horses, animals of the Central Asian steppes, adapted to the climate south of the Zagros Mountains. The Arabian Horse – the most noble horse-breed today – is the final result of this adaptation. This type of horses is already documented on the palace-reliefs at Niniveh, as well as some of the functions of horses in the Mesopotamian sphere. The commodities provided by the possession of the domestic horse range from milk, meat and skin to fast transportation and military strength. They also include a social function as status-symbol as well. In this respect only camels are comparable to the horses.

Up till now knowledge about the history of camel domestication is very scarce. The most reliable data comes from Mesopotamia where neither of the two camel species was at home. The wild two-humped camel survived in the desert areas of Central Asia. In antiquity its distribution reached west into the steppes east of the Caspian Sea, where its domestication may have taken place. The wild one-humped camel or dromedary originally occurred in the deserts of Arabia and North Africa. In Africa it became extinct during the Neolithic period.

Dromedary domestication took place in the Arabian part of its original habitat, but exact knowledge about when and where is still lacking. In south-east Arabia domestic dromedaries occur from the Iron Age onwards (Uerpmann and Uerpmann 2002, 2012). The obvious commodity provided by camels is their carrying capacity for riders and heavy loads.

The appearance of domestic camels in Mesopotamia is well documented in glyptic art. The most interesting find is a cylinder-seal which stylistically might derive

**Fig. 7.6** Imprint of a Mesopotamian cylinder-seal showing a two-humped camel ridden by two people sitting face-to-face on the two humps (Courtesy Prof. D. Potts)

from the middle of the 2nd millennium BC (Fig. 7.6). Recognition of the depicted animal is difficult, but obviously it is a two-humped camel. Apparently the artist who carved the seal had not seen the animal, because the humps are depicted as two cylinders. Two persons sit face-to-face on these cylinders on the back of the animal, which has the typical curved neck and the long legs of a camel.

Well dated glyptic evidence for domestic two-humped camels is found on the Black Obelisk of Shalmaneser III which was carved in the ninth century BC (Fig. 7.7). In both cases the artwork is outside of the natural geographic range of the two-humped camel. The same kind of evidence exists for the wild ancestor of the one-humped camel from within its natural range. Wild dromedaries are depicted on the monumental tombs of the Umm an-Nar culture in SE- Arabia together with other hunted animals (Fig. 7.8).

The frequency of dromedary bones in Bronze-Age contexts of SE-Arabia has sometimes been interpreted as indication of the domestic status of this animal. However, dromedaries react to domestication like all other large animals with a marked size-diminution during the initial stages of the domestication process. A clear diminution of body size is recognized between Late Bronze-Age and Iron-Age camel finds in SE-Arabia (e.g. M. and H.-P. Uerpmann 2012). In Mesopotamia the domestic dromedary is well documented on the Niniveh palace reliefs, where Arab warriors on camel-back (Fig. 7.9) fight against Mesopotamian soldiers on horses.

Fig. 7.7 Two-humped camels on the "Black Obelisk" of Shalmaneser III (ninth century BC)

Fig. 7.8 Reliefs of wild dromedaries on monumental Bronze Age graves in SE-Arabia

The glyptic evidence for camels in Mesopotamia is corroborated by bone-finds from a site in NE-Syria. The faunal remains from Tell Sheikh Hammad (Dur Katlimmu) provided bone finds of both camel species (Becker 2008). It is most interesting that they also indicate an earlier appearance of the Bactrian camel, followed by a later introduction of the domestic dromedary.

The last animal to be dealt with is the cat. It is a biological pest control, efficiently reducing mice and rats in stores and households. Thus it provides a com-

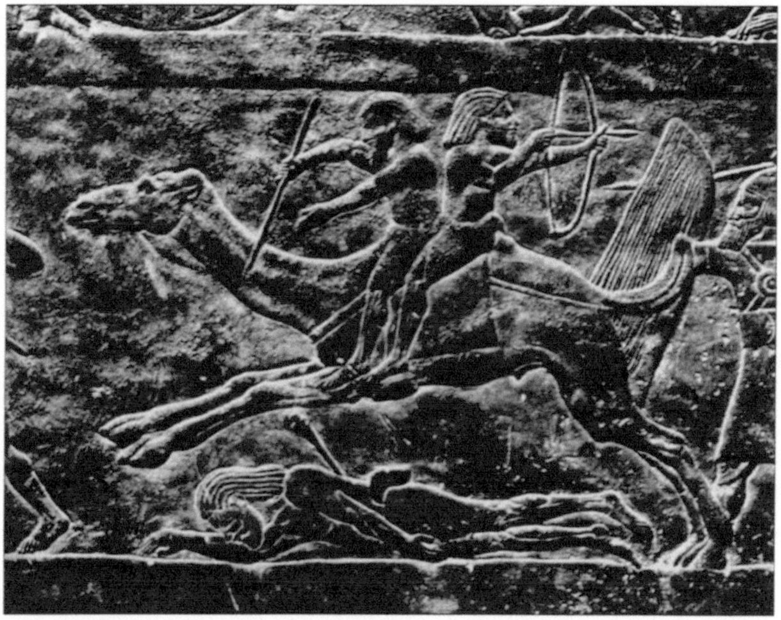

**Fig. 7.9** Arab warriors on camel-back depicted on the Niniveh palace reliefs

modity which is beyond the capabilities of the other animals described above. The cat may also have been introduced to Mesopotamia from outside, in this case presumably from NE-Africa. A date of its domestication cannot be fixed, because the process of cat domestication may have been discontinuous, and apparently it was not always induced by humans. To some extent cats seem to have domesticated themselves. They were attracted to settlements by the small rodents living there in high densities because of the year-round food storage in human houses. As pest-controls cats were accepted and gradually incorporated into human life as both useful and nice cute animal companions. An in-depth discussion of the related problems is found in Van Neer et al. (2014).

## 7.5 Conclusions

It can be summarized that the acquisition of animal-based commodities in SW-Asia started more than 10,000 years ago with the appearance of the domestic dog at EpiPaleolithic and Protoneolithic sites in the Levant. The sharp senses of dogs provided protection and help for the hunters. Not much later the domestication of sheep and goats initiated the transition from hunting to the controlled production of animal based food. The later domestication of pigs and cattle continued and diversified this development. In addition to meat and milk cattle also provided animal labor, a new category of commodity. Labor became the focus of later domestications. Asses, horses and camels supplied labor as their major contributions to human economy. Until the invention of modern energy management domestic animals remained the basic providers of this major commodity. Towards the end of the prehistoric period cats were a late addition to the domestic livestock in Southwest Asia providing some control over noxious rodents living on stored provisions but also paving the way for the later addition of luxury animals as a commodity for representative purposes and for providing joy.

## References


Becker, C. 2008. Die Tierknochenfunde aus Tall Seh Hamad/Dur-Katlimmu: Eine zoogeographisch-haustierkundliche Studie. In *Umwelt und Subsistenz der Assyrischen Stadt Dur-Katlimmu am unteren Habur*, Berichte der Ausgrabung von Tall Seh Hamad/Dur-Katlimmu 8, ed. H. Kühne, 61–132. Wiesbaden: Harrassowitz.

Bollongino, R., J. Burger, A. Powell, M. Mashkour, J–.D. Vigne, and M.G. Thomas. 2012. Modern taurine cattle descended from small number of near-eastern founders. *Molecular Biology and Evolution* 29(9): 2012–2104.

Clutton-Brock, J. 1999. *A natural history of domesticated mammals*. Cambridge: Cambridge University Press.

Conolly, J., S. Colledge, K. Dobney, J.G. Vigne, J. Peters, B. Stopp, K. Manning, and S. Shennan. 2011. Meta-analysis of zooarchaeological data from SW Asia and SE Europe provides insight into the origins and spread of animal husbandry. *Journal of Archaeological Science* 38: 538–545.

Darwin, Ch. 1868. *The variation of animals and plants under domestication*. London: John Murray.

Davis, S.J.M. 2005. Why domesticate food animals? Some zoo-archaeological evidence from the Levant. *Journal of Archaeological Science* 32: 1408–1416.

Dayan, T. 1994. Early domesticated dogs of the near east. *Journal of Archaeological Science* 21: 633–640.

Hongo, H., J. Pearson, B. Öksüz, and G. Ilgezdi. 2009. The process of ungulate domestication at Çayönü, southeastern Turkey: A multidisciplinary approach focusing on *Bos* sp. and *Cervus elaphus*. *Anthropozoologica* 44(1): 63–78.

Mellaart, J. 1967. *Çatal Hüyük: A neolithic town in Anatolia*. New York/London: Thames & Hudson.

Napierala, H., and H.-P. Uerpmann. 2010. A 'new' Paleolithic dog from central Europe. *International Journal of Osteoarchaeology* 22/2: 127–137.

Rossel, S., F. Marshall, J. Peters, T. Pilgram, M.D. Adams, and D. O'Connor. 2008. Domestication of the donkey: Timing, processes, and indicators. *Proceedings of the National Academy of Sciences of the United States of America* 105 (10): 3715–3720. [www.pnas.org].

Thalmann, O., B. Shapiro, P. Cui, V.J. Schuenemann, D.K. Sawyer, D.L. Greenfield, M.B. Germonpré, M.V. Sablin, F. Lopez-Caraldez, X. Domingo-Roura, H. Napierala, H.-P. Uerpmann, D.M. Loponte, A.A. Acosta, L. Giemsch, R.W. Schmitz, B. Worthington, J.E. Buikstra, A. Druzhkova, A.S. Graphodatski, N.D. Ovodov, N. Wahlberg, A.H. Friedman, R.M. Schweizer, K.-P. Koepfli, J.A. Leonard, M. Meyer, J.P. Krause, S.R.E. Pääbo, R.E. Green, and R.K. Wayne. 2013. Complete mitochondrial genome of ancient Canids suggest an European origin of domestic dogs. *Science* 342: 871–874.

Uerpmann, H.-P. 1987. *The ancient distribution of ungulate mammals in the middle east*, Beihefte zum Tübinger Atlas des Vorderen Orients. Wiesbaden: Dr. Ludwig Reichert.

Uerpmann, H.-P., and M. Uerpmann. 2002. The appearance of the domestic camel in SE- Arabia. *The Journal of Oman Studies* 12: 235–260.

Uerpmann, M., and H.-P. Uerpmann. 2010. Archaeozoology of camels in South Eastern Arabia. In *Camels in Asia and North Africa*, ed. E.-M. Knoll and P. Burger, 109–122. Wien: Verlag der Österreichischen Akademie der Wissenschaften.

Uerpmann, M., und H.-P. Uerpmann. 2012. Zug- und Lasttiere zwischen Maikop und Trialeti. In *Von Majkop bis Trialeti: Gewinnung und Verbreitung von Metallen und Obsidian in Kaukasien im 4.-2. Jt. v. Chr.*, eds. S. Hansen, A. Hauptmann, I. Motzenbäcker, und E. Pernicka (Hgb.), 237–251. Bonn: Habelt.

Van Neer, W., V. Linseele, R. Friedman, and B. De Cupere. 2014. More evidence for cat taming at the Predynastic elite cemetery of Hierakonpoplis (Upper Egypt). *Journal of Archaeological Science* 45: 103–111.

Weber, J.A. 2008. Elite equids: Redefining equid burials of the Mid- to late 3rd millennium BC from Umm el-Marra, Syria. In *Archaeozoology of the near east VIII, tome II*, ed. E. Vila et al., 499–519. Lyon: Maison de l'Orient et de la Méditerranée.

# Chapter 8
# The Beginnings of Metal Use in West Asia

**Ünsal Yalçın**

## 8.1 Introduction

Metal resources have played an important part in the development of human history. Therefore some periods of our cultural history have been named after certain metals, hence the Copper, Bronze and Iron Ages. Yet the invention of metals as raw materials and therefore their first utilization in the Old World goes back to the beginning of the Neolithic period. Here the oldest safely datable finds come from Anatolia and Northern Syria.

The tremendous development of the Neolithic cultures in the Old World, and the high value placed on copper as the first metal used there accelerated the development of metallurgy. This development happened in differentiated steps.

These steps in the development of metallurgy began in the Old World in different regions and different periods in time, but followed almost the same course. This phenomenon had technological, social and cultural reasons. The introduction of metallurgy depended on the demand of the societies for new materials and on the agricultural conditions and policies. Only members of a well-developed agricultural society paid attention to a valuable material like metal; and its high worth was obviously an important catalyst for its introduction (Strahm 1994).

This progression emerged from a background of general cultural change. Transformation was caused either by social differentiation or by the quick dispersal of cultural knowledge. The following is an attempt to establish the character and sequence of these steps based on present archaeological evidence. Each phase will be defined on the basis of technological development and compared with the accompanying social changes. In addition to technological criteria, cultural background

Ü. Yalçın (✉)
Deutsches Bergbau-Museum Bochum, Bochum 44791, Germany
e-mail: uensal.yalcin@bergbaumuseum.de



A. Tsuneki et al. (eds.), *Ancient West Asian Civilization*,
DOI 10.1007/978-981-10-0554-1_8

and the social context must also be considered. In this way the history of metallurgy can be integrated into a discussion of its social implications.

## 8.2 Human Use of Metals: History of Metallurgy

From the beginning of human history, raw materials have played an important role. Different stones and minerals were collected and used on various occasions. The colored minerals were preferred as pigments or were used to produce small objects like beads. This is particularly the case for iron oxides, which have been used for red coloring from the Paleolithic until today.

The first metal used by humans was native copper, which can be found in nature. Use of lead, silver, gold, antimony, arsenic, tin, zinc, iron and quicksilver followed native copper. These ten metals were the only metallic materials in use until the beginning of the industrial era in the eighteenth century AD.

Native copper was collected and the first objects, such as beads, were made from it. It was warm hammered at first to attain the favored shape. From ca. 5.000 BC copper was extracted from its ores. This is an important step in the development of metallurgy. Thus, after mastering extractive metallurgy people were able to produce copper. At the same time a new technique called casting came into use (Yalçın 2000a).

With the invention of extractive metallurgy people were able to control metal production. On the basis of experiments with different ores, different sorts of copper were produced. Step by step people learned to produce different types of copper. Thus they paved the way for alloy use, particularly bronze. Intermittently during the 3rd millennium BC and later more comprehensively in the second half of the 2nd millennium BC a new sort of metal called iron (and accordingly steel) displaced bronze. Finally in the 1st millennium BC zinc emerged as an important alloy component for producing brass and as well as quicksilver. Quicksilver was first used in medicine. These remained the only metallic materials used by humans until the eighteenth century AD.

The geology of Western Asia represents an asymmetric distribution of metallic raw materials. Two old continents, the African plate to the south and the Eurasian Plate to the north, collide with one another. In between them the Anatolian and Aegean micro continents were almost crushed. At this point the geological suture mountain chains such as the Carpathians, Taurus, Caucasus or Zagros rose, creating regions an abundant in rich metallic ores.

As a result of plate tectonics, regions like Anatolia and Northwest Iran have rich surficial areas next to metal ores, while other regions such as lower Mesopotamia do not have metal ores at all. Thus the ancient settlers of Anatolia, Northwest Iran, Caucasia and the Balkans had access to metals. Therefore it is not surprising that first metal finds occur in Anatolia, Northern Syria and Iran.

For this reason Anatolia can be considered the cradle of metallurgy. There has always been controversy about the leading role of Anatolia in the scholarly literature, although most of the safely datable finds investigated with the help of scientific methods came in fact from Anatolia (Esin 1976; Yalçın 1998a, 2000a, b; Yalçın 2008) and North Syria (Molist et al. 2009).

Focusing on Anatolia we can summarize the development of metallurgy in different stages. This model can be transferred to other regions where certain steps are evidently absent. Table 8.1 indicates the five stages of development in Anatolia:

1. Pre-Phase: The time without metal (Paleolithic-Neolithic: before 8.200 BC)
2. Beginning Phase: Single metal period in the Neolithic (since 8.200 BC)
3. Development Phase: Beginning of extractive metallurgy, Chalcolithic (approximately 5.000 BC)
4. Experimental Phase: Advanced metallurgy, Late Chalcolithic-EBA I (approximately 4.000 BC)
5. Industrial Phase: EBA II-Iron Age (after 2.800 BC)

If we compare Anatolia with other neighboring regions, we realize that the history of metallurgy and its development varied in different regions of the Old World. We can clearly observe that the earliest traces of metal use occurred in Anatolia and North Syria. However, there is also a highly visible commonality: Copper is the first metal of humans everywhere in West Asia.

Yet the development of metallurgy in Anatolia cannot be regarded in isolation. We can discuss this development only in conjunction with the rest of the Old World. If we observe the locations of archaeological finds in West Asia, we can state that metal was first used in the northern and southern regions at different times (Fig. 8.1).

## 8.3 The Neolithization and the Appearance of Copper as the First Metal

In the earliest human prehistory there is no use of metal. Its discovery and distribution as a raw material for arms and artefacts is one of the most important achievements of mankind. People have nonetheless used raw materials since very early times. Stone, bone and wooden artefacts were produced in order to facilitate food gathering or preparation. Hunters and gatherers appreciated the properties of special rocks or minerals such as flint and obsidian. They collected these, and used them to produce the first tools with cutting edges or drilling points. It is also possible that in observing nature and their environment, early human beings paid attention to colourful minerals and ores. Finally, they tried to use these objects for different purposes. In addition, we know that the origins of sedentary life can be located in the arc of southern Taurus that is in Eastern/Southeastern Anatolia.

By the end of the last ice age the conditions for sedentary life were optimal. The first settlements were founded in northwestern Iran, Northern Mesopotamia, the Levant, Eastern Anatolia and Cappadocia. In these regions the natural resources and the climate were suitable for the initial cultivation of plants and domestication of

**Table 8.1** Development stages of metallurgy in Anatolia (Yalçın 2008)

| Development- stage | Chronology | Technologies: mining, smelting, processing | Material/production | Sociocultural consequences |
|---|---|---|---|---|
| **Period I**<br>Pre-phase | Paleolithic-Neolithic (PPNA) >8200 BC | Collecting malachite and hematite; forming and handling of stone and obsidian | Coloured ore: used for pigment and jewellery (beads) | Organized collection and mining of obsidian and coloured ores |
| **Period II** | Neolithic (PPNB) since 8200 BC | Collecting malachite and native copper; annealing: beginning of pyrotechnology | Malachite and native copper: beads and other small objects | Raw material trade; workshops in the settlements |
| Beginning phase | | | | |
| | Neolithic (PN) since 7000 BC | Native copper: hot forging; manufacture of ceramics | Native copper, galena: beads and other small objects | Raw material trade; workshops in the settlement |
| **Period III**<br>Development Phase | Chalcolithic since 5000 BC | Near-surface mining; extractive metallurgy smelting in crucibles, casting | Oxide ores, smelting relics copper: standardized types, tools and jewellery | "Household metallurgy" raw material and metal trade; workshops in the settlements |
| **Period IV**<br>Building Phase | Late Chalcolithic-EBA I since 4000 BC | Underground mining; smelting in furnace | Complex and polymetallic ores; copper, arsenic copper, lead, silver, gold; standardized types | Extensive metallurgy in the workshops inside of settlements; raw material and metal trade |
| **Period V**<br>Industrial Phase | EBA II-Iron Age since 2800 BC | Centrally managed mining and metallurgy; smelting goes to the deposits processing/manufacturing in the settlements | Copper, lead, silver, gold, tin, bronze and other alloys; steel; since 1st mil.: brass serial production: ingots, weapons, tools, jewels | Centrally organized community with distinctive metal-culture supra-regional metal trade! emergence of regional states |

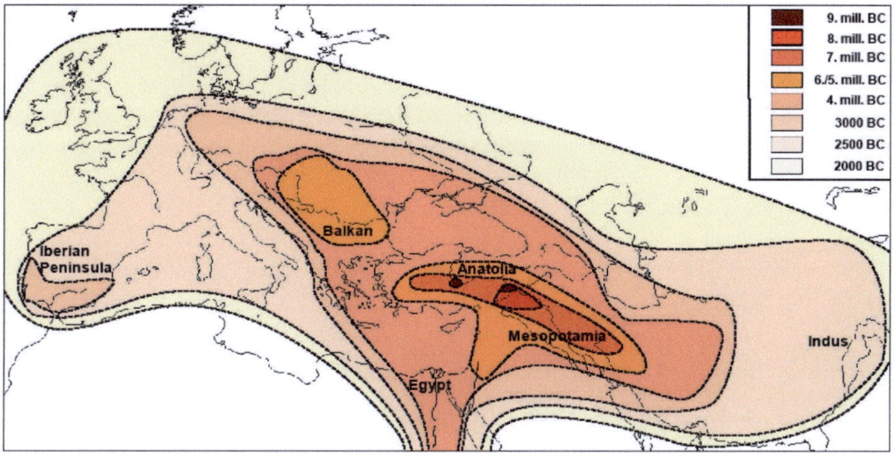


**Fig. 8.1** Metal diffusion

animals. This fact led to far-reaching changes in human life which can be observed from the developments at settlements like Hallan Çemi, Çayönü Tepesi, Nevalı Çori, Gritille, Cafer Höyük, Körtiktepe, Gusir Höyük and others in South-Eastern Anatolia, at Qermez Dere, Nemrik, and Zawi Chemi and Shanidar cave in North Iraq, at Tell Abr, Jerf el Ahmar, Tell M'laafat, Tell Halula and Mureybet in North Syria and at several sites like Jericho, Ain Mallaha or Ain Ghazal in the Levant (Aurenche 2007: 52, 53). Thus people left the hunting and gathering way of life to form sedentary and generative societies.

In this region the hunters and gatherers not only had an adequate access to wild animals for game, but they also had wild forms of cereals nourishing their diet (Bader 1989; Watkins 2007: 41–46). Later, from the PPNB on, they began the domestication of goat, sheep, cattle and pig. In the same period they also started to cultivate cereals and legumes. Thus the first productive societies emerged that were capable of holding animal stock. This is an important development in human history.

In the Neolithic Periods (PPNA-B) there was a small explosion in the use and extensive trade of status or luxury items. Among these, for example bead workshops (like Çayönü), plastered floors, and monumental buildings and imaginative statues (Nevali Çori, Göbekli Tepe, Gusir Höyük, Jerf el Ahmar) are all highly suggestive of the emergence of a social group controlling cult practices and also social and economic life (Aurenche 2007: 65; Hauptmann and Özdoğan 2007: 32–34).

A further sign of these changes is the increasing interest in densely coloured materials. At the pre-pottery settlements of West Asia, obsidian, hematite, carnelian, turquoise and malachite can be found almost everywhere. Like objects made of bone, antler and mollusc, they were carried as jewellery. As a result of experimental archaeology applied to stone and obsidian, scientists found out that the early settlers of Anatolia were able to produce small objects like beads from collected ores. In this

**Fig. 8.2** Oldest copper finds in West Asia

early phase of sedentary life green came into fashion. Green malachite and other copper minerals were collected in order to form small beads. The earliest examples come from Hallan Çemi and Çayönü Tepesi (Rosenberg 1994; Özdoğan and Özdoğan 1999). Green minerals were also used as make-up or for cosmetics.

Special attention should be given to the use of red in fashion before the beginning of sedentary life. In many Paleolithic and Epi-Paleolithic caves we find red predominating the palette. However at the beginnings of sedentary life we observe the emergence of green, therefore we can speak about green being in vogue then.

About 10.000 years ago in the pre-pottery Neolithic B (PPNB) curious collectors from Anatolia and North Mesopotamia brought native copper to their settlements and through experimentation the inhabitants were finally capable of working copper to produce small beads and other objects like awls.

In Anatolia more than a dozen copper mines are known to have yielded both oxide ores like malachite and native copper. In some of these mining regions, such as Ergani Maden, Kırmızı Tarla and Murgul, one can find pieces of native copper even today (Wagner et al. 1989; Yalçın 1998a). The only prehistoric native copper mine ever discovered is located in Derekutuğun, Northern Central Anatolia, there is (Yalçın and Maass 2013: 157, Fig. 4). So it is not by chance that the earliest copper finds come from these regions. North Iran probably has a similar area where copper ores rich in native copper could also be found.

In this context most of the secured dated Neolithic copper finds come from Anatolia (Fig. 8.2). The well-investigated evidence about this comes from Çayönü Tepesi (PPNB layers 2–6) and Aşıklı Höyük (PPNB layer 2). Pyrotechnology

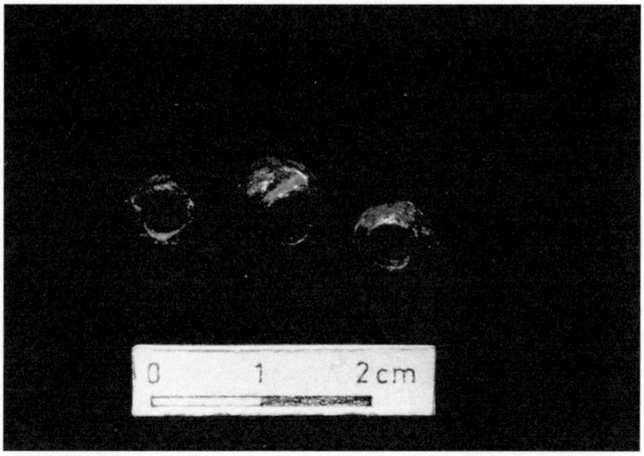


**Fig. 8.3** Copper beads from Aşıklı Höyük, PPNB (7800–7600 BC)

involving copper occurred there during a time without pottery (Özdoğan and Özdoğan 1999; Maddin et al. 1999). Most of the copper finds come from Çayönü (113 worked copper objects), and are composed of beads, pendants, and an awl and hook.

Burials in houses were discovered from the PPNB layers at Aşıklı Höyük. The dead were buried with copper beads (in the form of necklaces) as funeral gifts. More than 45 copper beads were found (Esin 1999). From the same period there is a copper bead from Nevalı Çori (e.g. Yalçın 2000a).

Investigations at Aşıklı Höyük point out that copper was formed by hammering. Copper was first hammered in thin sheets and the beads were rolled out of these sheets. In order to optimize the process the metal was heated (Figs. 8.3 and 8.4). Figure 8.4 illustrates the thin section of a copper bead from Aşıklı Höyük. In the microstructure various phases of recrystallization can be recognized. They show that the copper was heated several times and then it was hammered. Hammering distorts recrystallized grains. Thus pyrotechnology was one of the most important steps in human development. Pyrotechnology involving the working of copper occurred long before the firing of pottery. Even after the invention of pottery, native copper was the only choice for producing metal objects.

At another PPNB site called Tell Halula in North Syria 32 beads and a pendant were found (Fig. 8.5). These beads were also made from native copper (Molist et al. 2009). The microstructure of these objects show distorted recrystallized grains like those from Aşıklı Höyük and Çayönü. Here as well copper was warm hammered (Molist et al. 2009). Additionally three small rounded objects from Galena were discovered in Tell Halula. Two (uncertain and debatable) copper pieces were reported from Nemrik PPNB (Schoop 1995: 25; Ivanova 2013: 278/fn.16).

The techniques required for the development of ceramics were probably refined in the early PPNB within the context of using pyrotechnology to manipulate copper. The native copper was heated, enduring hammering until it glowed, in order to opti-

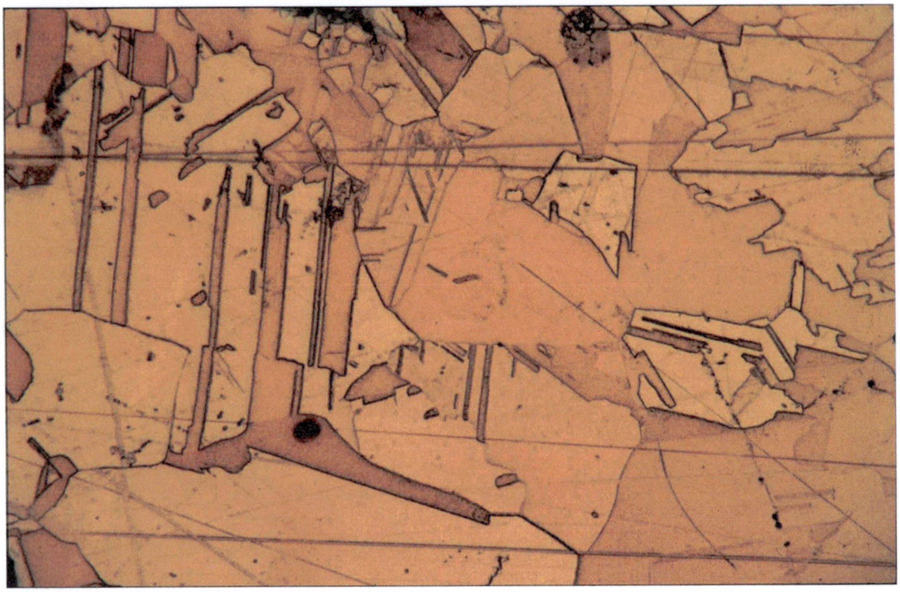

**Fig. 8.4** Micro structure of a copper bead from Aşıklı Höyük showing distorted and recrystallized grains indicating further heating and hammering

**Fig. 8.5** Copper pendant with small stone bead from Tell Halula (PPNB; from Molist et al. 2009: 38 fig. 4)

mize its shaping (Yalçın 2000a). This evidence proves that pyrotechnology in the processing of copper existed in Anatolia and northern Syria about 10.000 years ago. This was the first step on the long path of the development of metallurgy.

At the beginning of the PPNB other technological developments such as the burning of lime occur that are based on a pyrotechnical process (Hauptmann and Yalçın 2001). Lime was burnt in order to plaster the floors and walls of buildings. Many of the early settlements include communal buildings that were furnished with hard and burnished plastered floors (Fig. 8.6). Woolley has already drawn attention to this practice in his well-known work *Prehistory and the Beginnings of Civilization*

**Fig. 8.6** Çayönü Tepesi (PPNB). Terrazzo building with plastered floor

"**societies of plastered floor**" (Woolley 1958: 133–150). These lime or gypsum plastered floors are widely dispersed in the PPNB, but disappear in eastern Anatolia and the Levant with the invention of pottery, although continue to be used in Central and West Anatolia (Table 8.2).

Plastered skulls or masks are known from many Neolithic settlements (Table 8.2). Here either gypsum or lime was burnt and a mortar was produced. It is generally known that gypsum burns between 100 and 180 °C and lime needs a temperature above 800 °C. In this case a pyrotechnological method is obviously required.

Around 7.000 BC knowledge of pyrotechnology progressed further. At this time ceramics began to be fired and to assume an important role in daily life. In settlements of earlier periods vessels made from wood, leather, basketry and bark were widely used. In the Pottery Neolithic people started to form vessels from clay. Pots were heaped together and baked either in the open fire and pits or in simple household ovens.

During the Neolithic Period native copper remained the only metal in use. Finds dated to this time come from PN settlements at Çatalhöyük, Hacılar, Tepecik-Çiftlik (Fig. 8.7) and Canhasan (Fig. 8.8) in Central Anatolia, Tell el-Judeideh in Southeast Anatolia and other regions in West Asia. For example a flat pendant was excavated in Ramad (Israel) was dated to the beginning of the 7th millennium BC. It shows

**Table 8.2** Plastered floors in the Neolithic settlements in West Asia

| Site | Context | Date |
|---|---|---|
| Çayönü Tepesi | PPNB | 9. mill. BC |
| Nevalı Çori | PPNB | 9. mill. BC |
| Aşıklı Höyük | PPNB | 9./8. mill. BC |
| Musular | PPNB | 8. mill. BC |
| Suberde | PPNB | 8. mill. BC |
| Beisamoun | PPNB | 8. mill. BC |
| Yiftah El | PPNB | 8. mill. BC |
| Tell Eli | PPNB | 8. mill. BC |
| Nahal Oren | PPNB | 8. mill. BC |
| Munhata | PPNB | 8. mill. BC |
| Tell Farah | PPNB | 8. mill. BC |
| Hurbat Ghalil | PPNB | 8. mill. BC |
| Ain Ghazal | PPNB | 8. mill. BC |
| Ain Mallaha | PPNB | 8. mill. BC |
| Jericho | PPNB | 8. mill. BC |
| Beidha | PPNB | 8. mill. BC |
| Basta | PPNB | 8./7. mill. BC |
| Çatal Höyük | PPNB-PN | 8./7. mill. BC |
| Hacılar | PPNB-PN | 8./7. mill. BC |
| Ulucak | PN | 7. mill. BC |

**Fig. 8.7** Small beads (copper, stone and bone) from Niğde-Tepecik (ca. 6400 BC)

**Fig. 8.8** Mace head of Can Hasan 2b (ca. 6000 BC) is composed of native copper

**Fig. 8.9** Red painted bowl from Hacılar (ca. 6000 BC)

traces of cold hammering (France-Lanord and Contenson 1973). A small metal stick ("an awl") from Tell Maghzaliyeh (Iraq) (Ryndina and Yakhontova 1985), two copper beads from Tell Sotto (Iraq) (Merpert and Munchaev 1977; Bader 1989), and a bracelet and two pendants in a ring shape from Yarım Tepe (Iraq) (Merpert and Munchaev 1987) have been dated to the first quarter of the 7th millennium. Further finds come from Chaga Sefid (Iran) and Tell es-Sawwan (Iraq) (Schoop 1995: 28, 32).

Even the mace-head from Can Hasan (Fig. 8.8) which is dated to the Early Chalcolithic Period, appears to be a pseudo cast and was in fact formed by hammering (Yalçın 1998a).

In the Late Neolithic red dominates again. Ceramics are mostly red polished or with decorated with red (Fig. 8.9). Red pigments were used for the wall paintings in settlements like Çatalhöyük. Thus we can talk about red being in vogue again.

## 8.4 Later Development

Native copper remained the only metal for a further 3000 years. About 5.000 BC with the invention of extractive metallurgy, that is to say the smelting process, people began to extract metals in order to produce different sorts of metal. All over the Near East the extraction was done at settlements. Oxide copper ores were collected and brought to the settlement. At the workshops they were crushed and smelted in crucibles. Now people could produce metal when necessary. Later, as the surface ores were exploited, people started to go deeper and deeper following the ore veins; thus the in-ground reservoirs were mined. Only after the beginning of extractive metallurgy did metals other than copper finally influence daily lives. Thus metal turned out to be the most important raw material used for producing tools and weapons.

In the 3rd millennium (the Early Bronze Age II), the first big change in the development of metallurgy occurred (Period V). The ores were worked on an industrial scale near the mining areas (Strahm 1994; Yalçın 2000a, b, 2008). In the settlements, only evidence of further processing was found. These included crucibles,

**Fig. 8.10** Map of Anatolia showing EBA settlements of elites with mainly metal activities

**Fig. 8.11** Alacahöyük: Bronze bull figure with silver inlays

melting pots, ingots and manufactured objects (Müller-Karpe 1994). Well organized mining, metallurgy and the metal trade led to a society with hierarchic structures. With the beginning of the 2nd millennium BC the first regional kingdoms emerged that featured a central organized social structure and a distinctive metal culture.

Widespread socio-cultural changes also took place in the 3rd millennium BC. This can be traced through the dense metallurgic activities. Important centres of metal production and manufacturing arose. Well finished and highly artistic metal objects came to light in Troy, Alacahöyük, Horoztepe, Beycesultan, İkiztepe and in the other "royal residences" (Fig. 8.10) These are jewels with overwhelming beauty and precision, precious cult objects, weapons, tools and trade goods (Figs. 8.11 and 8.12).

Towards the end of the 4th millennium BC (the Early Bronze Age III) tin was used for the first time. In Tülintepe prestige objects made from copper were tinned (Yalçın and Yalçın 2009). Shortly after tin was alloyed with copper a new material called bronze was discovered. The discovery of bronze gave rise to the golden age of the metal trade. In this way tin was introduced as the essential element in bronze alloy.

**Fig. 8.12** Alacahöyük: Golden pot

In the Bronze Age this intensive metallurgy had an impact overall on human culture overall, because both the intense mining work and highly laborious metal work that followed required a well organised division of labour (Yalçin 1998b). The permanent labour supply required the release of miners, extractors and other craftsmen from other daily responsibilities. This fact led to a society with a division of labour and precisely defined social structures. The social elites generated a growing demand for metals and metal goods which in turn supported the increase in production. Mass production and interregional trade were the results of these relationships.

## 8.5 Conclusion

In conclusion we can state that the oldest metal finds from West Asia emerge in Anatolia and North Syria. From the 7th millennium BC on the number of metal finds increases. However copper remains the only metal in human use for a long time, until at the end of 5th millennium BC when new copper sorts and other metals like gold, silver and lead appear.

During the PPNB the first pyrotechnological products emerged. These are plastered floors, and skulls plastered with lime and gypsum. Native copper was warm hammered into small objects. These finds are part of an innovative development in human history, like a sedentary way of life, agriculture and religion.

Hence it is obvious that using raw materials, especially metals, was one of the essential steps in this development.

**Acknowledgments** I am indebted to my wife Dr. H. G. Yalçın (Ruhr Universität Bochum) and Prof. Dr. K. James-Chakraborty (School of Art History and Cultural Policy University College Dublin) for their assistance with English.

## References

Aurenche, O. 2007. Das "Goldene Dreieck" und die Anfänge des Neolithikums im Vorderen Orient, ed. Badisches Landesmuseum Karlsruhe, *Vor 12.000 Jahren in Anatolien. Die ältesten Monumente der Menschheit.* Katalog der Ausstellung: 50–53, 55, 56, 60–65.

Bader, N.O. 1989. *Earliest cultivators in Northern Mesopotamia*. Moskau: Academy of Sciences of the USSR, Institute of Archaeology.

Esin, U. 1976. Die Anfänge der Metallverwendung und Bearbeitung in Anatolien (ca. 7500–2000). Les Debuts de la Metallurgie, *Colloque* XXIII: 209–232.

Esin, U. 1999. Copper objects from the pre-pottery neolithic site of Aşıklı (Kızılkaya village, Province of Aksaray, Turkey). *The beginning of metallurgy*, ed. A. Hauptmann, Pernicka, Th Rehren, and Ü. Yalçın, Der Anschnitt, Beiheft 9: 23–30.

France-Lanord, A., and H. de Contenson. 1973. Une pendeloque en cuivre natif de Ramad. *Paléorient* 1: 107–115.

Hauptmann, H., und M. Özdoğan 2007. Die neolithische Revolution in Anatolien, . *Vor 12.000 Jahren in Anatolien. Die ältesten Monumente der Menschheit,* ed. Badisches Landesmuseum Karlsruhe, Katalog der Ausstellung: 26–36.

Hauptmann, A., and Ü. Yalçın. 2001. Lime plaster, cement and the first puzzolanic reaction. *Paléorient* 26(2): 61–68.

Ivanova, M. 2013. *The Black Sea and the early civilizations of Europe, the near east and Asia*. New York: Cambridge University Press.

Maddin, R., T. Stetch, and J. Muhly. 1999. Early metalworking at Çayönü Tepesi. *The beginning of metallurgy*, ed. A. Hauptmann, E. Pernicka, Th Rehren, and Ü. Yalçın, Der Anschnitt, Beiheft 9: 37–44.

Merpert, N.Y., and R.M. Munchaev. 1977. The earliest metallurgy in Mesopotamia. *Sovetskaya Arheologija* 3: 154–163.

Merpert, N.Y., and R.M. Munchaev. 1987. The earliest levels at Yarim Tepe I and Yarim Tepe II in Northern Iraq. *Iraq* 49: 1–36.

Molist, M., I. Montero-Ruiz, X. Clop, S. Rovira, E. Guerrero, and J. Anfruns. 2009. New metallurgical findings from the Pre-pottery Neolithic: Tell Halula (Euphrates valley, Syria). *Paléorient* 35(2): 33–48.

Müller-Karpe, A. 1994. *Altanatolisches metallhandwerk*, Offa-Bücher 75. Neumünster: Wachholtz.

Özdoğan, M., and A. Özdoğan. 1999. Archaeological evidence on the early metallurgy at Çayönü Tepesi. *The Beginning of Metallurgy*, ed. A. Hauptmann, E. Pernicka, Th. Rehren, and Ü. Yalçın, Der Anschnitt, Beiheft 9: 13–22.

Rosenberg, M. 1994. Hallan Çemi Tepesi: Some further observations concerning stratigraphy and material culture. *Anatolica* XX: 121–140.

Rydina, N.V., and L.K. Yakhontova. 1985. The earliest copper artifact from Mesopotamia. *Sovetskaya Arheologija* 2: 155–165.

Schoop, U.-D. 1995. *Die Geburt des Hephaistos. Technologie und Kulturgeschichte neolithischer Metallverwendung im Vorderen Orient*, Internationale Archäologie 24. Espelkamp: M.L. Leidorf.
Strahm, Ch. 1994. Die Anfänge der Metallurgie in Mitteleuropa. *Helvetia Archaeologica* 25: 2–39.
Wanger, G.A., F. Begemann, C. Eibner, J. Lutz, Ö. Öztunalı, E. Pernicka, and S. Schmitt-Strecker. 1989. Archäometallurgische Untersuchungen an Rohstoffquellen des frühen Kupfers Ostanatoliens. *Jahrbuch des Römisch-Germanischen Zentralmuseums* 36: 637–686.
Watkins, T. 2007. Der Naturraum in Anatolien. Ein Zusammenspiel von Klima, Umwelt und Ressourcen. *Vor 12.000 Jahren in Anatolien. Die ältesten Monumente der Menschheit*, ed. Badisches Landesmuseum Karlsruhe, Katalog der Ausstellung: 37–49.
Woolley, L. 1958. *History unearthed*. London: E. Benn.
Yalçın, Ü. 1998a. Der Keulenkopf von Can Hasan (TR): Naturwissenschaftliche Untersuchung und neue Interpretation. *Metallurgica Antiqua*, ed. Th. Rehren, A. Hauptmann, and J.D. Muhly, Der Anschnitt, Beiheft 8: 279–289.
Yalçın, Ü. 1998b. Frühe Eisenverwendung in Anatolien. *Istanbuler Mitteilungen* 48: 79–95.
Yalçın, Ü. 2000a. Anfänge der Metallverwendung in Anatolien. In *Anatolian Metal I*, ed. Ü. Yalçın, Der Anschnitt, Beiheft 13: 17–30.
Yalçın, Ü. 2000b. Frühchalkolitische Metallfunde von Mersin-Yumuktepe: Beginn der extraktiven Metallurgie?. *TÜBA-AR* 3: 111–130.
Yalçın, Ü. 2008. Ancient Metallurgy in Anatolia. *Ancient Mining in Turkey and the Eastern Mediterranean*, ed. Ü. Yalçın, H. Özbal, and A. G. Paşamehmetoğlu, International conference AMITEM 15–22 June 2008, Ankara: 15–42.
Yalçın, Ü., and A. Maass. 2013. Prähistorische Kupfergewinnung in Derekutuğun, Anatolien. In: *Anatolian Metal VI*, ed. Ü. Yalçın, Der Anschnitt, Beiheft 25: 153–194.
Yalçın, Ü., and H.G. Yalçın. 2009. Evidence for early use of tin at Tülintepe in Eastern Anatolia. *TÜBA-AR* 12: 123–142.

# Part III
# Urbanization and Change in Human Societies

# Chapter 9
# The Birth of Cities in Ancient West Asia

**Jason Ur**

## 9.1 Introduction

In what amounts to the blink of an eye in the history of humanity, we have become city dwellers. At the start of human existence, everyone moved; all human groups were hunters and foragers who moved seasonally, following the annual cycles of plants and animals. This pattern prevailed for millions of years. In West Asia, groups began staying in one place for some or all of the year only about 15,000 years ago. By 5000 years ago, cities had appeared. Today, the majority of humans on earth live in urban settlements. It was the most dramatic and rapid transition in human history. Where did it start, and how did it happen?

Before describing this history, we must define the subject. Today the "city" is largely a political designation. But we mostly think of cities as places with larger and denser populations than the other settlements around them (Cowgill 2004). Cities host activities and institutions that exist only in them, and not in smaller nearby settlements; for example, cities are centers of government, places of industry and manufacture, homes to powerful religious institutions and the gods themselves. These institutions allow for communities to organize themselves not by kinship, but through other class, religious, or economic relationships independent of family connections. Modern definitions stress one thing that cities do *not* do: sustain themselves. The specialists in government, crafts, and religion do not produce the food that they require to live. This has been an important aspect of urban definitions: cities can only exist with *hinterlands*. They do not exist as islands; they need smaller agricultural villages around them, to supply them with basic food, and they in turn supply the villages with things they cannot produce themselves: manufactured goods, or services like government or connections to the divine through temples.

J. Ur (✉)
Department of Anthropology, Harvard University, 11 Divinity Avenue,
Cambridge, Massachusetts, 02138 USA
e-mail: jasonur@fas.harvard.edu



A. Tsuneki et al. (eds.), *Ancient West Asian Civilization*,
DOI 10.1007/978-981-10-0554-1_9

There are, however, some less functional aspects to cities. Cities are meaningful places. Great events might have happened at a city, or it may be known to be especially sacred, even the home of a divine being. Such symbolic aspects are difficult to find in the earliest cities, but it is safe to say that in addition to these functional characteristics, many people come to cities, and stay in them, because they feel that urban life is the proper way to exist in the world, in proper alignment with society and with the cosmos. These aspects are certainly accurate for cities of the past few millennia, but whether they describe the earliest cities is still an open question.

After such a long and successful existence as mobile foragers, and a shorter but equally successful history of village agriculture, why, then, did our ancestors start to come together? In many traditional societies, villages never grow to cities because internal disputes split them apart (Bintliff 1999). Within the village, separation is the only way to solve such disputes. Most scholars of urban origins in ancient West Asia see cities as an adaptation, a new settlement form that was deliberately created because it would solve problems, like the ones that resulted in splitting. For some, cities were an invention to benefit all members of society, through more efficient access to both food and specialized products and services (for example, Nissen 1988). Others think cities were a device to protect the interests of a new class of elites, and therefore urban life mostly benefited only *some* members of society (Adams 1981).

This review will consider the origins of cities in West Asia. It will take a critical look at the urban definitions used by archaeologists, and it will question the functional understandings of urban origins. It will include a new theory of the birth of cities that sees them originally as accidents, unintended consequences of social changes. In this understanding, cities did not originate as an adaptation for either some or all of society; cities as functionally specialized places was a later development, not the impetus for urban origins.

Ancient Mesopotamia was the land between the Tigris and Euphrates Rivers (Fig. 9.1). Today this region encompasses nearly all of Iraq, and large parts of Syria and Turkey. At its northwestern areas, the rivers are deep, and flow in narrow floodplains, but with adjacent broad plains that were very productive for agriculture. To the southeast, the rivers run across the southern Mesopotamian plain. This region is so flat that the rivers cannot carry silts and clays. They have to drop them and then they rise up atop their own levees. The elevation of the rivers makes small-scale irrigation an easy undertaking. For many scholars, this geographical advantage explained the long term durability of cities in southern Mesopotamia (Algaze 2008).

For a long time, archaeologists assumed that cities originated in the south of Mesopotamia, and that the first city was Uruk, which originated around 3200 BC. Recent research, however, now shows that the first city-like settlements took their form in the *north* of Mesopotamia, starting about 1000 years earlier.

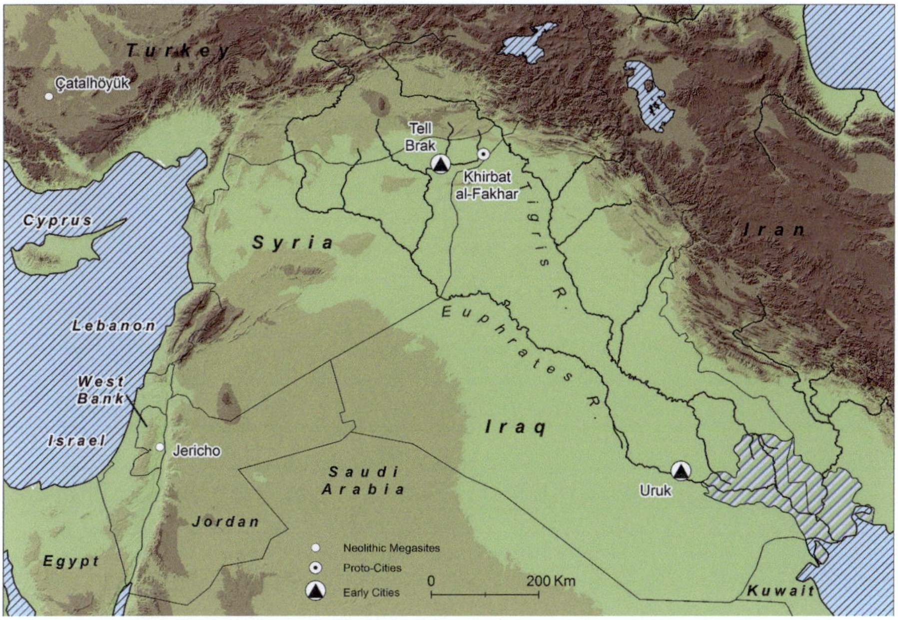


**Fig. 9.1** West Asia, with sites mentioned in the text

## 9.2 The Challenges of Early Cities to Traditional Archaeological Methods

By their very nature, early cities present huge challenges to archaeological research. To put it very simply, cities are large, and excavation, which is the traditional method for collecting data, does not work well with large phenomena. Excavation techniques are excellent for recovering information about a feature or a small set of features: for example, a tomb, or the rooms of a domestic structure. To excavate an entire neighborhood takes an investment of years, if not decades. To excavate an entire village might take an entire career. Many lifetimes would be required to dig a city completely. These challenges are merely the ones related to time; just as prohibitive are the vast sums of money that would be required.

To approach early cities in a holistic manner, archaeologists have developed several tools that allow us to investigate large areas quickly. These methods return different kinds of information than traditional excavation. Two of the most successful new techniques are satellite imagery analysis and archaeological survey.

### *9.2.1 Satellite Remote Sensing and Early Cities*

The human view of the earth is hopelessly tethered to the ground. Our own experience with cities shows us that at any moment, we can only perceive a small portion of them, excepting some views from the tallest skyscrapers. But from a vertical

**Fig. 9.2** CORONA satellite photograph of Tell Brak, northeastern Syria (taken 11 December 1969)

perspective, we can visualize entire cities at once. This remote perspective often reveals elements of size and structure that would be very difficult to perceive on the ground (Kouchoukos 2001). Two examples are illustrative.

Tell Brak was an early city in northeastern Syria. Today it is a mass of collapsed mudbrick that rises 40 m over the plain (Oates et al. 2007). It is incorrect to call it *one* ancient city; it was *several* ancient cities, each built in a different form on top of its predecessor (Ur 2014b). Excavations have given us important windows into parts of the city in various eras. A vertical perspective, however, opens a new window. In a US CORONA spy satellite image, Brak's mound is clearly visible (Fig. 9.2). Beyond the mound, however, the fields are discolored, especially to the northwest and northeast, signaling former areas of settlement that did not grow into great mounds. Such areas are difficult to perceive on the ground without close observation. This image shows that Tell Brak was about five times larger than just its high mound.

Remote sensing can also inform about impacts of cities beyond their edges and walls. At Brak, the settlement area is surrounded by a series of dark radiating lines. These lines are areas where vegetation grows especially thickly, or where water

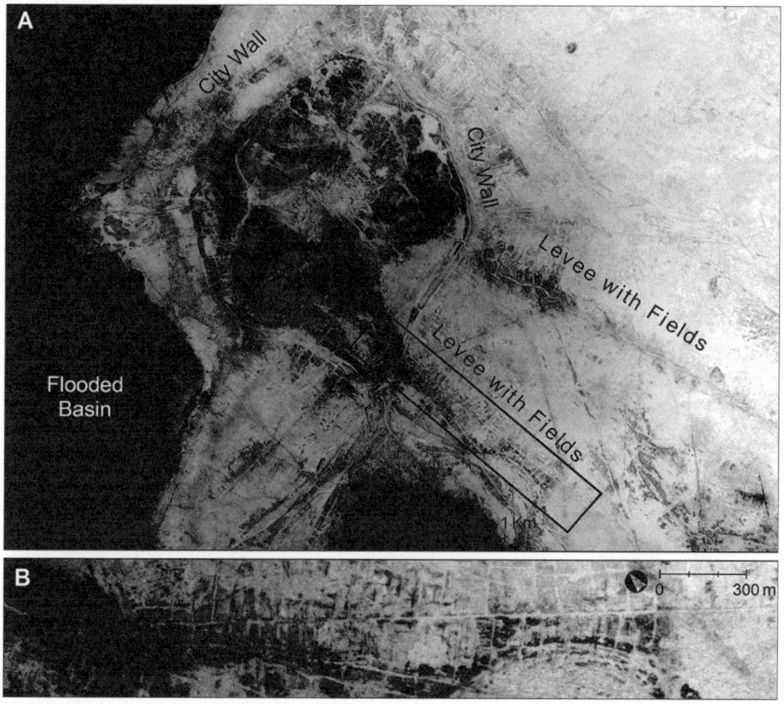


**Fig. 9.3** CORONA satellite photograph of Uruk, Iraq (taken 4 May 1968). (**a**). Uruk with city wall and ancient watercourses; (**b**). detail of ancient watercourse and fields

accumulates, or both. Measured on the photographs, they are up to 100 m wide. Most of these features are invisible on the ground, but a few are slightly depressed. These photographs have captured ancient paths of movement, tracks that extended from the city out into its fields, to the pastureland beyond, and to neighboring towns and villages.

CORONA photographs can reveal the agricultural systems that supported ancient cities as well. For the second example of remote sensing, consider the famous city of Uruk in southern Iraq, which is frequently invoked as "the world's first city" (although see below for an alternate view). Uruk's city wall is enormous, and clearly visible on CORONA photographs (Fig. 9.3a), but archaeologists have known of it for a century. Newly revealed in this image is the elaborate system of ancient irrigation that surrounded the city. To its southeast, for example, two ancient watercourses ran away from the city (Fig. 9.3b). These watercourses may have originated as natural features, but the urban farmers of Uruk used them for irrigation. The CORONA image preserves the details of the fields in an elaborate network. These ancient fields have since been destroyed by the expansion of modern agriculture; they survive only on CORONA photographs.

Historic intelligence photographs, such as CORONA, captured the landscape in black and white film, but modern satellite imagery can show color, including light

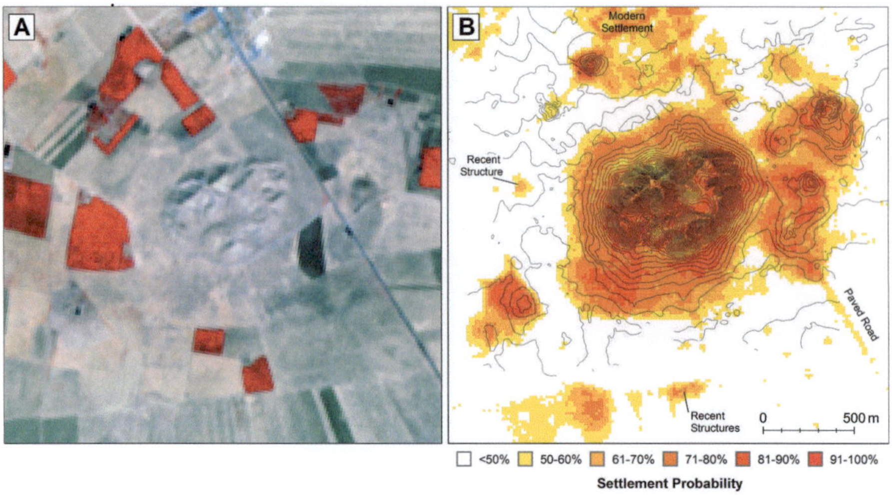


**Fig. 9.4** Satellite remote sensing at Tell Brak, Syria. (**a**). ASTER satellite scene (acquired 31 August 2004); (**b**). ASTER-derived settlement probability map (Menze and Ur 2012)

in wavelengths that the human eye cannot see. Such "false color" images can see in a broad range of the electromagnetic spectrum. In West Asia, the ASTER sensor has been especially useful. ASTER is a cooperative effort between the US and Japan that is very strong in the near infrared, and can see into the thermal infrared part of the spectrum. Because ASTER produces digital images, its data can be analyzed in an automated fashion by computer algorithms. Automated ASTER analysis has been applied to Tell Brak (Fig. 9.4a). By training computer software to investigate hundreds of ASTER images, it is possible to search for the combinations of pixels that the computer thinks represent the "signature" of archaeological sites in the electromagnetic spectrum. The results give the probability that a given place is an archaeological site. ASTER images of Tell Brak show the central mound, and the fields surrounding it. An ASTER analysis examined over 30 images of Brak, and produced a probability map (Fig. 9.4b).

### *9.2.2 Archaeological Survey and Early Cities*

These satellite remote sensing methods can suggest *where*, but they do a poor job of telling us *when* an ancient site was occupied. To know *when*, archaeologists must visit places on the ground, and employ the techniques of archaeological survey. Survey makes surface observations of archaeological sites. In west Asia, sites have three distinguishing features (Fig. 9.5). Sites are mounded. Centuries or millennia of inhabitation in mud brick houses causes sites to rise up, and to decay into artificial hills called *tells* once they are abandoned. Sites have a high density of artifacts on their surfaces. Settlement debris comes to the surface in many ways, especially

**Fig. 9.5** The three main criteria for archaeological sites in West Asia: moundedness (*left*), high artifact density (*center*), and anthropogenic soil discoloration (*right*)

through erosion and agricultural plowing. Finally, their soils are discolored, compared to the natural soils of the Mesopotamian plains. This last feature is what makes them visible in satellite imagery.

Archaeological survey is the most comprehensive approach to early Mesopotamian cities. Ancient people disposed of their household debris in the streets in front of their houses. Survey methods assume that where one finds artifacts of a certain age, people lived in that part of a site during that age. By making systematic collections of artifacts across the surface of a site, archaeologists map its shifting patterns of settlement.

Archaeological survey has been especially successful at Tell Brak. Since the 1970s, archaeologists from the University of Cambridge have excavated Brak's central mound. Because Brak's earliest history is so deeply buried, they have only small windows into its origins, including its patterns of growth. Furthermore, they knew that the fields around the mound were full of artifacts. The site was much larger than the central mound, but how large was unclear. For this reason, the team conducted a systematic archaeological survey of Brak's outer city (Fig. 9.6a). They made collections of surface artifacts at systematic intervals of 50 m, in most parts of the site, guided by hand-held GPS receivers. At each collection area, all pottery that was chronologically significant was collected from a ten by ten meter square, counted and weighed. Ultimately almost a thousand collections were made across the site.

Artifacts were all washed and analyzed to determine how old they were. For ceramics, the style of production changed over the course of centuries, as potters changed their techniques and adopted new decorative styles. Ceramics can, therefore, be dated visually, by considering their style and manufacturing techniques. Most time periods have distinctive "diagnostic" types.

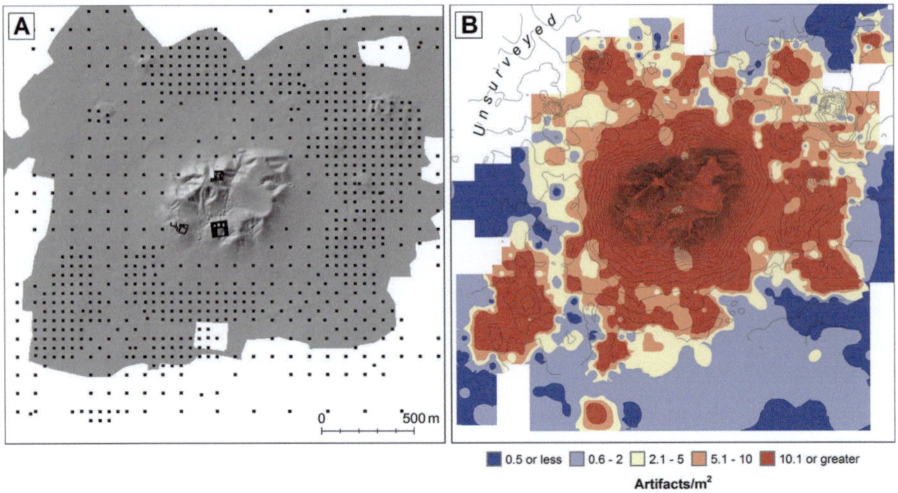


**Fig. 9.6** Field survey at Tell Brak, Syria. (**a**). Site topography, excavation areas, and survey collection units; (**b**). Surface artifact density

When we map the total density of artifacts, the central mound is revealed to be a small part of the overall site; most of the city is today under the fields around it. When this map of artifact density (Fig. 9.6b) is compared to the site probability map (see above, Fig. 9.4b), the similarities are remarkable. In the case of Tell Brak, analysis the surface artifacts and the soil discoloration produce nearly identical results. Most importantly, both show an enormous site, and surprisingly early in time, as will be discussed below.

## 9.3 Early Cities and Non-Cities in West Asia

Before looking at the cities revealed by these techniques, let us consider a few famous early places in the west Asia that are *not* cities.

### *9.3.1 Early Experiments and "Proto-Cities"*

The earliest candidates for urban status are often the so-called "Neolithic Megasites," most famously Jericho in Palestine, and Çatal Höyük in central Anatolia. They have some characteristics associated with cities. Jericho has a monumental wall surrounding it with a high tower. Çatal Höyük had domestic structures packed together at a high density that we associate with cities. It also had new and elaborate forms of ritual. Both of these sites were large for their time, but both lack most other

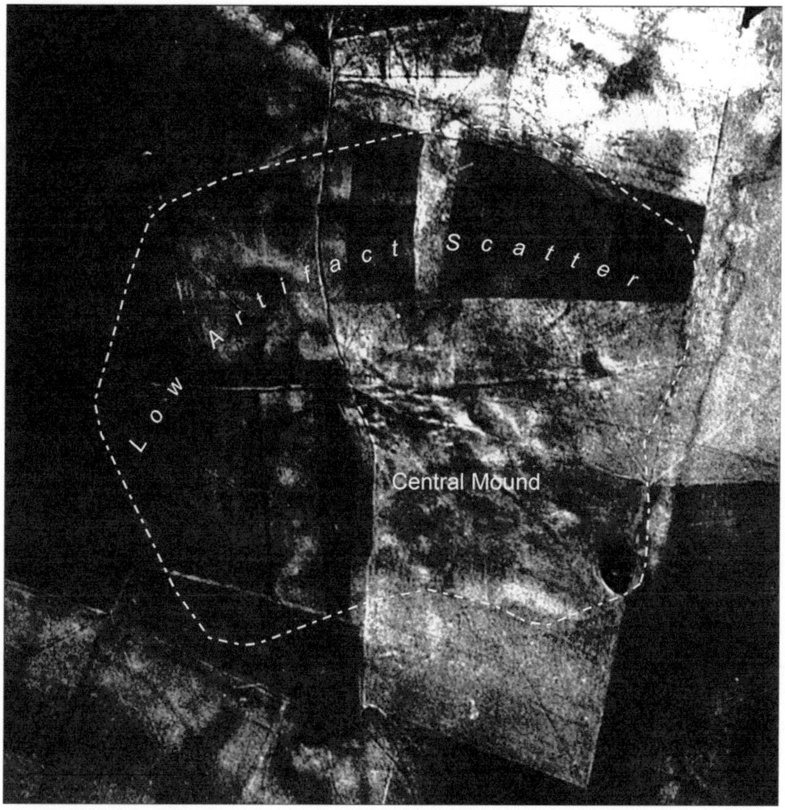


**Fig. 9.7** The proto-city at Khirbat al-Fakhar (Hamoukar southern extension), Syria. CORONA image taken 11 December 1969

characteristics of cities. They appear to have been experiments with nucleated settlement, but experiments that did not develop further, especially in the realm of population (Emberling 2003).

Another early experiment is found in northeastern Syria at Khirbat al-Fakhar, to be dated to the end of the fifth millennium BC (Ur 2010; Al Quntar et al. 2011). Khirbat al-Fakhar (also called the "southern extension" of Hamoukar) has a central mound of 31 ha that would be, by later standards, a large town (Fig. 9.7). It is surrounded, however, by a vast low area of surface artifacts, including huge amounts of obsidian. This fine stone had been brought to the settlement from hundreds of kilometers away, and evidence for manufacture of tools is found throughout. In total, the site covers some 300 ha, or thirty times the size of Jericho or Çatal Höyük. It does, however, appear to have been low density. On CORONA satellite imagery, the discoloration of the outer low areas is discontinuous, which suggests that houses or neighborhoods had been separated by open spaces.

Khirbat al-Fakhar appears, therefore, to have been another early experiment in new settlement form. Something brought people together, probably trade in obsidian. These people did not, however, feel comfortable living in close proximity as they did at Çatal Höyük. Apparently, there did not exist social mechanisms to permit them to live together closely, and it is possible that some of the inhabitants were only present on a seasonal basis. It also appears that the settlement did not last very long, maybe only a few generations. In sum, we might consider Khirbat al-Fakhar to be a sort of "proto-city" (Al Quntar et al. 2011).

Only a few centuries later, and a few hundred kilometers to the west, another early experiment took place at Tell Brak. Like Khirbat al-Fakhar, the earliest large settlement was composed of a central mound with dispersed outlying neighborhoods. We know this from the surface collection described earlier, which showed artifacts from this time period in a dispersed pattern. The central mound was fully settled, but it was surrounded by isolated small neighborhoods that were close to it, but still separated by hundreds of meters of open space (Ur et al. 2007; Ur et al. 2011). As at Khirbat al-Fakhar, something drew people to this settlement, but could not integrate them. The social separation of these outlying communities was manifested by spatial separation. In total, this discontinuous settlement at Tell Brak covered 55 ha.

Excavations on the mound give us a sense of what new social forms may have developed (Oates et al. 2007). For example, one building had exceptionally large walls, and an enormous basalt threshold. Nearby was found a drinking chalice formed of obsidian, marble, and bitumen—all exotic foreign materials. Anyone walking across that threshold would have known that this was an important residence, and anyone seen drinking from such a chalice would be marking themselves as an important individual.

### *9.3.2 Emergence of High-Density Early Cities*

By the middle of the fourth millennium BC, Brak had expanded. The outer neighborhoods enlarged and grew together. The scatter of artifacts is so dense that it suggests not only growth in area, but also greater density of occupation. Brak now covered approximately 130 ha of dense settlement, more than ten times the size of any of its fourth millennium neighbors (Ur 2014b). Excavations on the mound show that new social institutions were forming, probably related to religion. The Eye Temple was a large and elaborately decorated building with a long central hall. Within it were found thousands of small "eye idols," which probably represent worshippers. The building itself sat atop a large artificial terrace. We know of no other contemporary institutions like the Eye Temple, and it is possible that it shows us the creation of a new institution that gained its authority through a privileged connection to the gods (Emberling 2002).

It is likely that Brak's urban growth, and the formation of new social institutions, was not a smooth process. These changes may have necessitated, or caused, a great deal of violence within the city. Evidence comes from an area on the northern edge of the city, where human bodies were cast out with the garbage, and then a feast was consumed (McMahon et al. 2011; Liverani 2006). Excavations have revealed dozens of unburied bodies amidst common settlement debris, but with particularly rich animal remains atop them. With no neighbor of comparable size, it is difficult to view these people as an attacking army. Rather, they are likely to have been the losers in an internal struggle within the city, the victims of the growing pains of one of the first cities in the world.

At the end of the fourth millennium, we finally come to the great city of Uruk. The recent discoveries at Khirbat al-Fakhar and Tell Brak show that Uruk can no longer be viewed as the locus of urban origins, but rather the culmination of a series of increasingly successful experiments in settlement nucleation over more than a millennium. At Uruk by around 3100 BC, the world finally witnessed a settlement that appears to have all of our preferred urban characteristics (Liverani 2006; Crüsemann et al. 2013). Uruk was huge: 250 ha, about twice the scale of Brak a few centuries earlier. It was certainly a center of religious and political power. At its center were huge and well-decorated buildings that must have marked the households of political elites. Nearby was another tripartite building, the White Temple. It was similar to the Eye Temple at Brak, in that it sat atop a high terrace—likely to be the household of a divine being.

## 9.4 The Birth of Cities: How and Why

Brak and Uruk are often assumed to have attained urban status because of a "revolution" in which society shifted from kinship basis to new non-kin forms of social relations, perhaps based on residence or class (Childe 1950; Adams 1966). Alternatively, it is possible that what happened was not the *replacement* of kinship, but rather an evolution of how kinship was defined, particularly the nature of the household. From later periods, we know that temples and palaces were run like households, and using household terminology like father, son, and servant (Gelb 1979; Schloen 2001). In the course of the fourth millennium, households were not replaced by "the state," but rather became something larger than just an immediate family. Households were able to contain other households, to the extent that entire settlements, or even cities, were conceptualized as very large metaphorical households (Ur 2012, 2014a).

To describe this social transition better, we must go backwards, before Uruk, Brak, or even Khirbat al-Fakhar. Let us consider the houses of the fifth-millennium Ubaid period, a time of village settlement before the appearance of cities at Brak and Uruk. Ubaid houses had a long central hallway with small rooms on either side (Aurenche 1981). The central hall was the location of many household activities, most importantly, communal activities like eating meals (Roaf 1989). The Ubaid village at Tell Abada in Iraq contained many such houses. The largest houses had central halls of 3 m width, and 10 m long (Fig. 9.8a).

Now consider the Eye Temple at Tell Brak from the middle of the fourth millennium. It has all of the spatial characteristics of the house at Tell Madhhur and elsewhere (Fig. 9.8b). In other words, it looks like a house. But it was elaborately

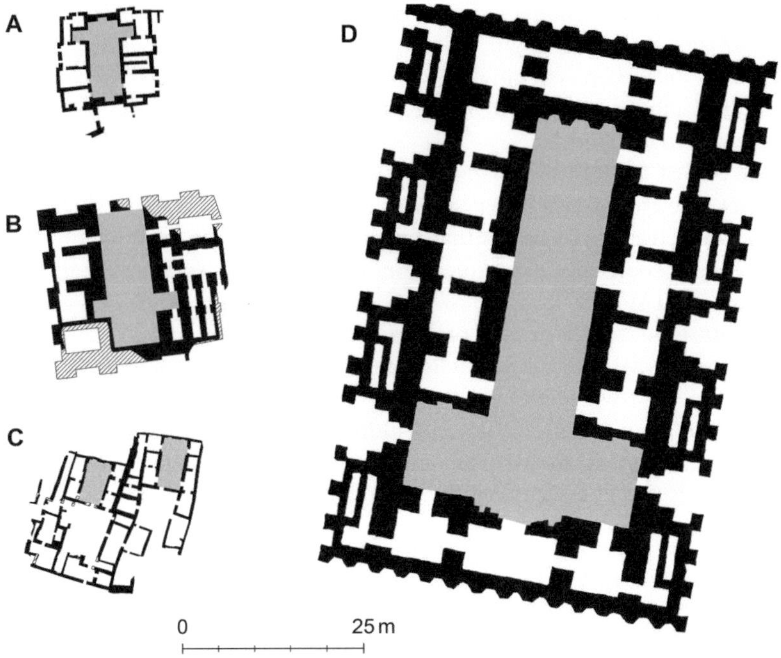


**Fig. 9.8** Ubaid and Uruk central hall houses (central halls in gray). (**a**). Ubaid house at Tell Madhdhur. (**b**). Uruk period "Eye Temple" at Tell Brak; (**c**). Uruk houses at Hamoukar; (**d**). Temple D at Uruk

decorated with mosaic walls, was full of religious artifacts, and sat atop a huge terrace. Archaeologists call it a "temple," but to the people of Brak, it was another household—this time the household of a god. Indeed, unambiguous private houses of the time look identical, for example the houses at Hamoukar (Fig. 9.8c)

Finally we come to Uruk at the end of the fourth millennium. Its central district had many such tripartite buildings, of even larger size and greater elaboration. At the most extreme end of the scale was Building D, which had a central hall 55 m long and over 10 m wide (Fig. 9.8d). It is important to note, however, that the difference between the houses of the Ubaid period, and the structures of Uruk over a 1000 years later was not of kind, it was of degree—all share the same ground plan. A resident of Uruk might leave his own home and enter one of these massive buildings and be very impressed, probably overwhelmed with awe, but he would definitely recognize that he was in a house writ large; different from his own in scale, but still easily interpreted by him as a house.

Thus the so-called "urban revolution" was not revolutionary at all. The new social institutions, whether divinely focused temples or secular palaces, were not really new, but rather a reformulation of a very familiar social institution: the household. Some household heads we label as "priests," but they were the servants of the divine head of household. Others we call "kings" but they were only the patriarchal heads of metaphorical households that encompassed entire towns and cities. The demographical shifts to nucleation seem revolutionary to us, but to the ancient inhabitants, it was just an expanded form of their own households and therefore familiar and acceptable, in a way that real revolutionary change never is.

Finally, it may be that these first cities, at Brak and Uruk, may have been quite accidental. If the alternative model of urban origins is correct, these cities can be envisioned as unintended consequences of the reinterpretation of the household. The sociologist Max Weber described a common social form called patrimonialism, in which societies were conceived as households, and their political economies were directed almost solely toward satisfying the needs of the household (Weber 1978). He used ancient Mesopotamia as an example, and nearly a century of further scholarship in the historical eras has proven him largely correct. It is easy to imagine that a successful and ambitious head of a household would actively try to expand its membership by social connection to others, as we can document for the third millennium BC. This expanding household would have spatial consequences; it would draw in new followers. The intention of the household head (i.e., the man we label as "king") was *not* to form a city, it was to sustain and expand his household. But the unintended consequence of that action would be the growth of this settlement. In later times, after the city as a settlement form had emerged, it could be copied deliberately. These first durable cities at Brak and Uruk, however, may not have been foreseen by their earliest rulers. The initial stages of this process are found in the low density or de-centralized spatial patterns at Khirbat al-Fakhar, and its final stages in the great palace and temple households at Uruk (Ur 2014a).

This new social model is admittedly hypothetical. It needs more research, and explicit testing with carefully targeted excavation. Unfortunately, most of ancient Mesopotamia is in dark times at present. Syria is in the midst of a civil war, and there are reports of fighting between Kurds and fundamentalist Islamic fighters on the mound of Brak itself. This sectarian conflict has now spread to Iraq. Tragically, it may be a long time before we can return to these questions in the field. But we must return; these questions are fundamental to who we are as a global urban society.

**Acknowledgements** I am indebted to the Directorate General of Antiquities and Museums in Damascus for permission to carry out research and Brak and Khirbat al-Fakhar/Hamoukar. The Tell Brak Suburban Survey was co-directed by Philip Karsgaard, with survey assistants Fahid Juma'a and Shilan Ramadan, and benefited from the advice and support of Joan and David Oates, Geoff Emberling, Henry Wright, Eric Rupley and Carlo Colantoni. Funding for the Suburban Survey was provided by the British Academy, the McDonald Institute for Archaeological Research (University of Cambridge), the University of Michigan and Harvard University. CORONA imagery is reproduced courtesy of the United States Geological Survey.

## References


Adams, R.M. 1966. *The evolution of urban society*. Chicago/New York: Aldine.

Adams, R.M. 1981. *Heartland of cities*. Chicago: University of Chicago Press.

Al Quntar, S., L. Khalidi, and J.A. Ur. 2011. Proto-urbanism in the late 5th millennium BC: Survey and excavations at Khirbat al-Fakhar/Hamoukar, Northeast Syria. *Paléorient* 37: 151–175.

Algaze, G. 2008. *Ancient Mesopotamia at the dawn of civilization*. Chicago/London: University of Chicago.

Algaze, G. 2013. The end of prehistory and the Uruk period. In *The Sumerian world*, ed. H. Crawford, 68–94. Oxford/New York: Routledge.

Aurenche, O. 1981. *La maison orientale, l'architecture du Proche-Orient ancien des origines au milieu du quatième millénaire*. Paris: Paul Geuthner.

Bintliff, J. 1999. Settlement and territory. In *Companion encyclopedia of archaeology*, vol. 1, ed. G. Barker, 505–544. London: Routledge.

Childe, V.G. 1950. The urban revolution. *Town Planning Review* 21: 3–17.

Cowgill, G.L. 2004. Origins and development of urbanism: Archaeological perspectives. *Annual Review of Anthropology* 33: 525–549.

Crüsemann, N., M. van Ess, M. Hilgert, and B. Salje (eds.). 2013. *Uruk: 5000 Jahre Megacity*. Peterberg: Reiss-Engelhorn-Museen, Deutsches Archäologisches Institut.

Emberling, G. 2002. Political control in an early state: The eye temple and the Uruk expansion in northern Mesopotamia. In *Of pots and plans: Papers on the archaeology and history of Mesopotamia and Syria presented to David Oates in honour of his 75th birthday*, ed. L. al-Gailani Werr, J. Curtis, H. Martin, A. McMahon, and J. Reade, 82–90. London: Nabu Publications.

Emberling, G. 2003. Urban social transformations and the problem of the 'First City': New research from Mesopotamia. In *The social construction of ancient cities*, ed. M.L. Smith, 254–268. Washington, DC/London: Smithsonian Institution.

Gelb, I.J. 1979. Household and family in early Mesopotamia. In *State and temple economy in the ancient near east*, ed. E. Lipinski, 1–97. Leuven: Departement Oriëntalistiek, Katholieke Universiteit Leuven.

Kouchoukos, N. 2001. Satellite images and near eastern landscapes. *Near Eastern Archaeology* 64: 80–91.

Liverani, M. 2006. *Uruk: The first city*. London/Oakville: Equinox.

McMahon, A., A. Sołtysiak, and J. Weber. 2011. Late chalcolithic mass graves at Tell Brak, Syria, and violent conflict during the growth of early city-states. *Journal of Field Archaeology* 36: 201–220.

Menze, B.H., and J.A. Ur. 2012. Mapping patterns of long-term settlement in northern Mesopotamia at a large scale. *Proceedings of the National Academy of Sciences* 109: E778–E787.

Nissen, H.J. 1988. *The early history of the ancient near east, 9000–2000 BC*. Chicago: University of Chicago.

Oates, J., A. McMahon, P. Karsgaard, S. al-Quntar, and J. Ur. 2007. Early Mesopotamian urbanism: A new view from the north. *Antiquity* 81: 585–600.

Pollock, S. 1999. *Ancient Mesopotamia: The Eden the Never Was*. Cambridge: Cambridge University Press.

Roaf, M. 1989. Social organization and social activities at Tell Madhhur. In *Upon this foundation: The 'Ubaid' reconsidered*, ed. E.F. Henrickson and I. Thuesen, 91–146. Copenhagen: Carsten Niebuhr Institute of Ancient Near Eastern Studies.

Schloen, J.D. 2001. *The house of the father as fact and symbol: Patrimonialism in Ugarit and the ancient near east*, Studies in the archaeology and history of the Levant 2. Winona Lake: Eisenbrauns.

Ur, J.A. 2010. *Urbanism and cultural landscapes in northeastern Syria: The Tell Hamoukar survey, 1999–2001*, Oriental Institute Publications 137. Chicago: University of Chicago Oriental Institute.

Ur, J.A. 2012. Southern Mesopotamia. In *A companion to the archaeology of the ancient near east*, ed. D.T. Potts, 533–555. Malden/Oxford: Blackwell.
Ur, J.A. 2014a. Households and the emergence of cities in ancient Mesopotamia. *Cambridge Archaeological Journal* 24: 249–268.
Ur, J.A. 2014b. Urban form at Tell Brak across three millennia. In: *Preludes to urbanism: The late chalcolithic of Mesopotamia*, eds. A. McMahon, and H. Crawford. Cambridge: McDonald Institute for Archaeological Research.
Ur, J.A., P. Karsgaard, and J. Oates. 2007. Urban development in the ancient near east. *Science* 317: 1188.
Ur, J.A., P. Karsgaard, and J. Oates. 2011. The spatial dimensions of early Mesopotamian urbanism: The Tell Brak suburban survey, 2003–2006. *Iraq* 73: 1–19.
Weber, M. 1978. *Economy and society*, vol. 2. Berkeley: University of California.

## Further Reading

There are few works on urban origins intended for the general reader. Recent reviews of the transition to urbanism in ancient West Asia include Algaze (2008, 2013), Pollock (1999), and Ur (2012).

# Chapter 10
# The Invention of Cuneiform and the Dawn of the Modern World

**Jonathan Taylor**

## 10.1 Introduction

We are very unusual human beings. For maybe 97 % of modern man's time on this planet, no-one anywhere ever read or wrote anything at all. Yet today we live in a world where writing is everywhere. It is so much a part of life that we don't realise just how much we depend on it. Literacy is considered a fundamental, democratic right; illiteracy meanwhile is considered a profound social disadvantage, and often a stigma. Even a couple of hundred years ago, the situation was very different.

"History Begins at Sumer," as a leading Sumerologist once famously said (Kramer 1961).[1] The earliest writing we know anywhere in the world comes from Iraq over 5000 years ago. Around 3200 BC an accountant in the temples of Uruk, in southern Iraq, invented a system that allowed him to keep track of the movement of goods in high detail. It was not the first time that anyone had kept track of such information. There was a long history in the region of using a range of clay tokens in different shapes; they must have represented animals or quantities of goods. Groups of these tokens were gathered into a clay envelope, having first pressed them into its surface (thus providing a record on the outside of what is inside the envelope). The surface of the envelope was impressed with another new invention—the cylinder seal. Cylinder seals are carved with a design in reverse so that

[1] It is important to keep in mind the distinction between something being the first of its kind and it being the first step on the way to the modern equivalent. The connections are usually more subtle and indirect. Cuneiform is not directly ancestral to any currently used writing system, for example. Some scholars argue that it inspired ancient Egyptian writing, which in turn inspired the alphabet. Mesopotamian cuneiform is fascinating to us because it provides a wealth of evidence detailing some of the first efforts of Man to struggle with issues we still face today. Charpin 2010 offers an interesting overview of what literacy meant in Mesopotamia.

J. Taylor (✉)
British Museum, London WC1B 3DG, United Kingdom
e-mail: JJTaylor@britishmuseum.org



A. Tsuneki et al. (eds.), *Ancient West Asian Civilization*,
DOI 10.1007/978-981-10-0554-1_10

when rolled onto clay, a scene appears (see Collon 2005); this functioned as a kind of signature or office stamp.

What made writing special was the precision and power of the new system. It had an almost magical ability to record and transmit information, quickly, effectively and accurately. The accountant could never have imagined just how much his invention would change the world. Once the dark veil of prehistory was lifted, an ever greater library of texts brings us closer to understanding the thoughts and dreams of our ancient forebears, their hopes and fears, their lives and deaths. The latest securely dated tablets come from the first century AD, by which time its use was restricted to scholarly, particularly astronomical, purposes; thus Mesopotamian cuneiform writing lasted 3500 years. The time that separates those last astronomers from the origins of their writing system is twice as much as that which separates them from us today. Put another way, cuneiform documents fully two thirds of recorded human history.

## 10.2 First Writing

The first documents display ever more sophisticated formatting schemes (see Nissen et al. (1993) for an accessible introduction to early bookkeeping). These explain the role of each piece of information on the tablet. Multiple transactions could be gathered in a single space, without losing the detail of the constituent parts. At this time, characters were grouped into boxes. Writing in the order in which you read the characters would come only later. The first writing is a system made up of 1000 or more characters, plus multiple different counting systems, depending on what kind of thing you were counting. Many of the characters were pictures of (parts of) things in the real world, such as a stalk of barley or a jug filled with beer (see Fig. 10.1a and b).[2] Beer was a staple commodity drunk daily by men, women and children

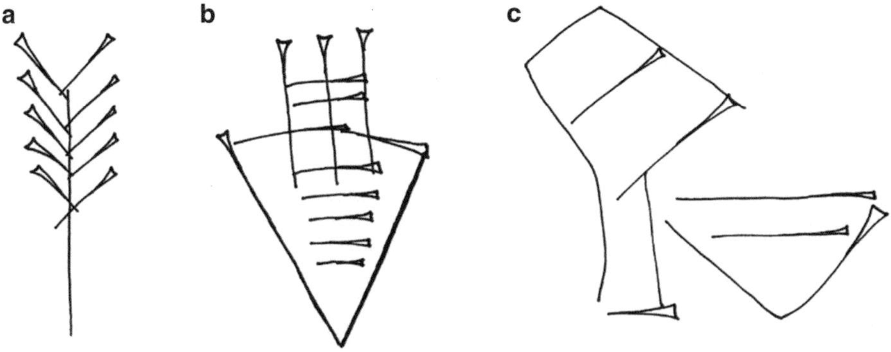


**Fig. 10.1** Some of the key elements of the first writing: (**a**) BARLEY; (**b**) BEER; (**c**) RATIONS

[2] See http://www.britishmuseum.org/research/collection_online/collection_object_details.aspx?objectId=327218&partId=1 for an example of a tablet bearing these signs.

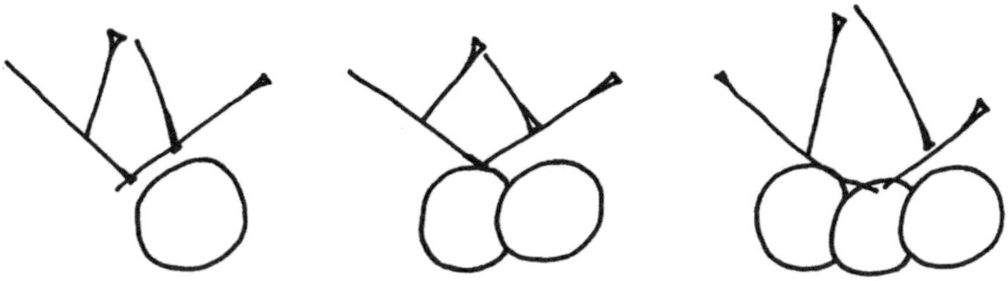

**Fig. 10.2** The characters DAY 1, DAY 2 and DAY 3 from BM 116730, recording transactions over 5 days

alike. Water was not clean. Beer, meanwhile, had been treated, and was a lot safer. The writing system was a tool designed to allow bureaucrats to maintain complex accounting in large temple households, managing large numbers of workers. A character that appears often on the tablets is a human head and what seems to be a picture of a special type of mass produced bowl, which contains just enough food to keep a man alive for a day (see Fig. 10.1c). This character indicates that the tablet deals with rations for workers. The quantities of barley being issued can be very large indeed. There are numbers for quantities that are enough to feed someone for a month, 6 months, 5 years, 15 years and even 150 years. Of course these quantities were used to record rations issued not for a few people over a very long period, but rather for large numbers of people over a short period.

Writing mastered space. There are characters for the many cities of Sumer, for example, and their major institutions. More significantly, writing also mastered time. There are characters to indicate day, month and year. DAY is a picture of the sun rising in the east. Combining DAY with numbers lets you write documents that record not only a single transaction but a week's worth, for example. Tablet BM 116730 illustrates this well.[3] It contains five rows of text, each separated from the next by a double line. Thus there are five different pieces of information. Along the left hand side is the character for "day" plus a number. This is day 1, day 2, day 3 and so on (see Fig. 10.2). MONTH and YEAR work similarly. Scribes could easily track transactions over days, months and years.

Remarkably, it is difficult to say for sure what language these first scribes were writing. What we have is not flowing prose, but terse administrative notes. When a few centuries later we are able to determine what language is being written, in a reformed version of this writing, that language is Sumerian. Sumerian has no known relations. Our understanding of it is still coloured by our grasp of Akkadian, the second main language which was to be written in cuneiform.

There are various stimuli that may have encouraged the invention of writing in Mesopotamia. One is the complexity of society, although it is possible to have very complex societies or big cities with no writing. Another is technology. The admin-

[3]The original tablet can be seen at http://www.britishmuseum.org/research/collection_online/collection_object_details.aspx?objectId=327216&partId=1.

istrative system using tokens and sealings, and markings on clay, may have inspired writing. It is possible to run a very advanced administration with tokens and sealings, but writing is a much more powerful tool. Thirdly there are linguistic arguments for why writing should have been invented here rather than somewhere else. Why ever it was that writing was invented, it constitutes intellectually a giant leap for mankind. It fundamentally changes the world.

## 10.3 How Cuneiform Works

Although sometimes criticised today as complicated and difficult to learn, cuneiform is a wonderfully flexible kind of writing, which enjoyed a career of 3500 years and was used to write more than a dozen different languages, many of them unrelated (see Finkel and Taylor 2015 for an introduction to cuneiform). The following examples give a flavour of how it works. First is a stylised drawing of a human head (see Fig. 10.3a), such as seen in Fig. 10.1 above. It represents a person. The two main languages written in cuneiform in Mesopotamia were Sumerian and Akkadian. Signs potentially can be read in either language. Thus HEAD, for example, can be read directly as the word *sag* in Sumerian or *reshum* in Akkadian. HEAD can also be used to provide just the sound sag or *resh*; Akkadian makes frequent used of signs in this way, using sounds derived from either Sumerian or Akkadian.

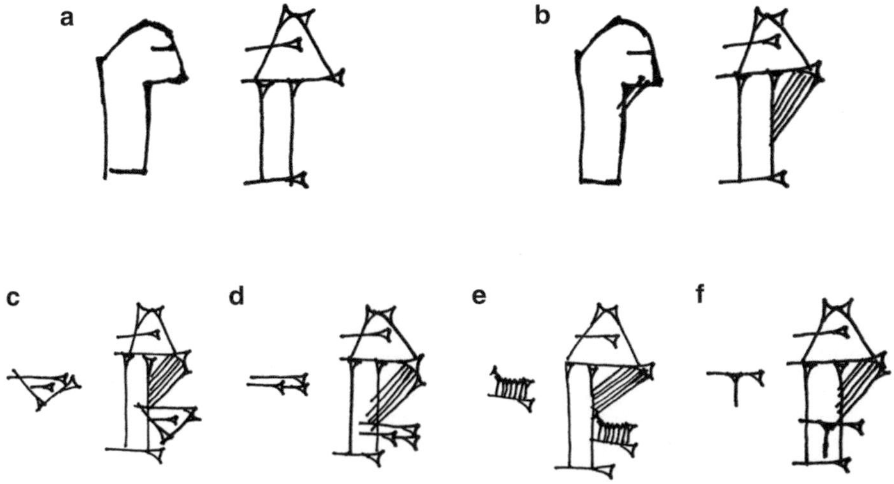


**Fig. 10.3** *Top row:* (**a**) the character HEAD in its early and later versions; (**b**) the character MOUTH in its early and later versions; *Bottom row:* (**c**) the character BREAD (originally the ration bowl seen in Fig. 10.1) when placed inside MOUTH means "to eat"; (**d**) WATER (originally two wavy lines) placed inside MOUTH means "to drink"; (**e**) MOUTH plus HAND means "to pray"; (**f**) MOUTH plus *me* is read *eme* "tongue."

Some signs are derived from other signs. Shading the mouth area of HEAD produces the sign MOUTH (see Fig. 10.3b). MOUTH can be used to write the word "mouth," but also other words related to it, such as "voice" and "word"; any sign can potentially have several readings. MOUTH can be further developed. Adding the sign BREAD (originating from the ration bowl seen in Fig. 10.1c above) gives EAT (see Fig. 10.3c). Adding instead WATER gives DRINK (see Fig. 10.3d). More sophisticated ideas can be communicated just as easily. Adding HAND to MOUTH gives PRAY (based on a gesture of prayer; see Fig. 10.3e). There's also a charming character meaning "secret," which is written with MOUTH with a box across it, looking somewhat like a zip. Alternatively, helping signs can be added to show how to read the main sign; for example, MOUTH plus *me* must be read *eme* "tongue" (see Fig. 10.3f).

As mentioned, MOUTH can be used to write "voice." This is *gu* in Sumerian. Nowadays, we call this "*gu*-three." This is because of another principle of cuneiform: there can be different signs for writing a particular sound. Thus there is also a "*gu*-one" (meaning "thread"), "*gu*-two" (meaning "bank (of a waterway)"), "*gu*-four" (meaning "ox"), as well as various other *gu*s. There may have been differences in how these words sounded in Sumerian—some scholars have suggested that Sumerian may have been a tonal language, like Chinese—but Sumerian has no known relatives to help us. We reconstruct it partly through Akkadian; we don't really know what Akkadian sounded like either, even though it is related to Semitic languages spoken today, such as Arabic and Hebrew.

During the third millennium BC, the individual strokes used to write cuneiform became less curved, and the signs become more abstracted in appearance. The strokes also took on a more wedge-like appearance—hence the name cuneiform, which means simply "wedge-shaped."

## 10.4 The Rich Cuneiform Library

Cuneiform was used across the Middle East, particularly Iraq, but also at times in parts of Iran, Armenia, Turkey and the Levant. Even the Egyptians at one point used cuneiform for their international correspondence. Cuneiform writing was a tool created not for religious purposes, nor for literature, nor in the service of kings, nor even for shopping lists, but to serve accountants. That remained its primary use throughout its history. While the vast majority of cuneiform texts are administrative or legal in nature, it was not long before new uses were found for writing. It had a remarkable ability to record and transmit information, quickly, effectively and accurately. Only a couple of hundred years after the earliest traces of writing, we begin to see the world's first literature, myths, hymns and prayers. We see the names and deeds of kings recorded, laws, international treaties, letters, magic, medicine, recipes for food and industrial production, mathematics, astronomy and astrology. The cuneiform corpus tells us what it means to be human in another time and another place, which helps us understand what it means to be human here and now.

It has been estimated that there are as many as half a million cuneiform tablets in the world. That's only a fraction of what was originally produced. The tablets we have do not give us an even picture of the whole history of the cuneiform world. Some times and places are documented in glorious detail, while others languish in obscurity. Someone once counted the scribes known to us from a large group of texts dating to around 2000 BC, a period known to us as one of intensive bureaucratic activity (maybe 1/5 of all known tablets date to that time). There are about 2000 scribes named in those texts. If we imagine that each was active for around 10 years, worked about 200 days per year, and produced one document per day, then we reach a combined output of 4 million tablets. What does that really mean? If we were to arrange them like books on a shelf, that shelf would be 100 km long, stretching all the way from Tokyo to Mount Fuji and beyond. Travelling on a Shinkansen it would take 45 min to reach the last tablet on the shelf. And that's just the scribes we know about, from a period that covers only a century (and mostly within not much more than a single generation). If we then consider the long career of cuneiform, even assuming much lower rates of production at all other times we quickly reach over 40 million tablets; that would take seven and a half hours to journey along by high speed train. The numbers of specialists in the world who can pick up a tablet and read it is so small that they could all join you on that train, and there would still be empty seats. That makes studying cuneiform a magical experience. We read tablets that nobody else has read for thousands of years, and there's plenty more left to read. For anyone who is excited at the prospect of making new discoveries, cuneiform is an ideal area of study.

Why do tablets, made from simple river mud, survive over thousands of years? Very few cuneiform documents were designed to last. Most were not baked hard to preserve them, but simply left to dry in the air. They were mostly ephemeral records, surviving accidentally having been thrown away in antiquity. They survive amazingly well unless they come into contact with water. Millions remain in the ground, awaiting discovery by future generations. Mesopotamian archaeology is characterised by the excavation of "tells." Tells are essentially man-made hills that rose from the flat countryside as a result of generations of re-building mudbrick structures. Discarded tablets were treated in the same way as broken mud-bricks. They are routinely found as building debris, in floors, walls, staircases, and sometimes even in bricks themselves. We are lucky that in the environment of Mesopotamia these clay tablets survive rather well, despite the rough treatment they received.

Among the highlights of cuneiform are what may be the world's first attempt to compile a comprehensive library. The seventh century BC library of King Ashurbanipal of Assyria may have inspired the famous library of Alexandria. Both libraries were burnt down, but whereas this destroyed Alexandria's texts, it preserved Ashurbanipal's clay tablets as crisp and clear as the day they were written. Some of those tablets tell us of other types of document that were lost: there were once many writing boards of wood or ivory filled with wax. None of those survive from the library. A couple of tablets also bear fragmentary inscriptions in ink. In the last days of cuneiform, under the influence of other writing systems, scribes must sometimes have written in ink. Almost no trace of that survives; one wonders how much has been lost. The destruction of Ashurbanipal's library saw many of its tablets broken into pieces. Modern scholars delight in re-joining these fragments, rather like doing a jigsaw puzzle. In the case of the library, it's like doing about 10,000 jigsaw puzzles all at once.

## 10.5 Clay and Reed and the Human Hand

Cuneiform writing technology was closely tied to the Mesopotamian environment. There was too little rain in the south to support agriculture, so the cities lined up along the two rivers that give Mesopotamia its (modern) name ("land between the rivers," that is, the rivers Tigris and Euphrates). These rivers fed a network of canals, letting the citizens grow cereals and vegetables. From the floods of these rivers came the clay on which writing was put. In the rivers grew reeds, from which styli were cut. Cuneiform writing was a craftsman's tool. It could be used effectively enough by less skilled scribes, but in the hands of masters it was exquisitely balanced and beautiful.

Cuneiform writing was controlled. There was a set order in which the strokes of each sign were made. This must have been one of the first things a Mesopotamian schoolboy learnt. This order is visible through distortions in the surface of the clay tablet. Clay from a wedge can be pushed into the hole left by a previous wedge, or the straight edge of a wedge can be bent by a subsequent wedge. Thus we learn rules such as that the wedges were made left to right, bottom to top, container sign before contents, and various others.

We can see how tablets were made from looking at broken examples. The scribes used different qualities of clay for different purposes. Some are very fine, with an appearance almost like porcelain. Others contain small stones, or sometimes plant matter or mollusc shells. This means that tablets can also be valuable sources of information about the ancient environment. The clay would be formed into tablets of different shapes and sizes depending on the type of text about to be written. Making a tablet was a skilled task. The simplest method was just to hand mould clay. A well known type of tablet produced in ancient schools is a small round one that fits within the palm of the hand. They were made by patting the hands together or slamming a ball of clay against a flat surface.

Another technique was to take pieces of clay and push them together in rows, like in a bar of chocolate. Alternatively, a sheet of clay was rolled over itself and then squashed flat. A more sophisticated, and probably more usual, method was to take two sheets of clay. The first was folded over itself. The result was placed on the second sheet, at ninety degrees, and this was folded over the first. Clay envelopes were applied in a similar way; envelopes were invented in Mesopotamia in the third millennium BC as a way to protect documents.

## 10.6 Twenty-First Century Mesopotamian Expertise

Mesopotamian experts were famous in antiquity for their medical skills, their ability to interpret omens, and their wealth of astronomical observations. Their astronomy has left an indelible imprint on the modern world (for Mesopotamian astronomy/astrology, see Hunger and Pingree 1999). When a rocket is launched into space, when someone navigates with GPS or surveys an archaeological site with satellite imaging, it's based on the fact that a circle has 360°. There's absolutely no reason why a circle should not have a 100°. That it has 360 is ultimately because the Babylonians used sexagesimal maths in their astronomy (for Mesopotamian maths in its own context, see Robson 2008). Similarly, every time you look at a watch, you are practicing Babylonian maths. There's no reason why there should not be 100 min in an hour, but again, there are 60 because of the Babylonians.

## References

Collon, D. 2005. *First impressions: Cylinder seals in the ancient near east*. London: British Museum Press.
Finkel, I., and J. Taylor. 2015. *Cuneiform*. London: British Museum Press.
Hunger, H., and D. Pingree. 1999. *Astral sciences in Mesopotamia*. Leiden/Boston: Brill.
Kramer, S.N. 1961. *History begins at Sumer*, 2nd ed. London: Thames & Hudson.
Nissen, H., P. Damerow, and R. Englund. 1993. *Archaic bookkeeping. Writing and techniques of economic administration in the ancient near east*. Trans. P. Larsen. Chicago/London: University of Chicago Press.
Robson, E. 2008. *Mathematics in ancient Iraq: A social history*. Princeton: Princeton University Press.

## *Further Reading*

Charpin, D. 2010. *Reading and writing in Babylon*. Cambridge, MA/London: Harvard University Press.
Crawford, H. 2013. *The Sumerian world*. Abingdon: Routledge.
Leick, G. 2007. *The Babylonian world*. Abingdon: Routledge.
Radner, K., and E. Robson. 2011. *The Oxford handbook of Cuneiform culture*. Oxford/New York: Oxford University Press.
Van de Mieroop, M. 2004. *A history of the ancient near east, ca. 3000–323 BC*. Oxford: Blackwell Publishing.

# Chapter 11
# The World's Oldest Writing in Mesopotamia and the Japanese Writing System

**Jun Ikeda and Shigeo Yamada**

## 11.1 Invention of Writing and Its Prehistory

Agriculture and animal husbandry were first carried out about 8000 BC in Syria-Palestine and northern Mesopotamia, where dry farming was possible. These new life skills were introduced to southern Mesopotamia about 5000 BC, when agriculture was made possible by the use of irrigation technology. After that, however, southern Mesopotamian culture assumed the leading role in Mesopotamian civilization. Toward the end of the fourth millennium BC, the city – that is, the concentration of population in association with large buildings and a fortification system – came into existence in southern Mesopotamia. The world's oldest writing system was invented in the city of Uruk in about 3200 BC as a new means of recording and communication. This first writing in human history is attested by clay tablets inscribed with proto-cuneiform (linear-pictographic) signs discovered from the temple complex at Uruk. The documents were various administrative records related to the city, where people of different occupations and classes lived under a hierarchy headed by a ruler and temple authority and many kinds of goods and products were apparently assembled, sorted and distributed. Slightly later, from the very end of the fourth millennium BC onward, word lists, including occupations and the names of cities, were also made alongside the administrative texts. Writing then spread rapidly to the cities of southern Mesopotamia surrounding Uruk, such as Adab, Umma, Jamdat-Nasr near Kish, and then further to a broader area in Mesopotamia and Syria.

The prehistory of writing also has gradually been clarified. Archaeological investigation has revealed that small clay objects, called "tokens," were continuously used in villages and towns in Mesopotamia and its surroundings from c. 8000 BC

J. Ikeda • S. Yamada (✉)
Faculty of Humanities and Social Sciences, University of Tsukuba,
1-1-1, Tennodai, Tsukuba 305-8571, Ibaraki, Japan
e-mail: yamada.shigeo.gu@u.tsukuba.ac.jp



A. Tsuneki et al. (eds.), *Ancient West Asian Civilization*,
DOI 10.1007/978-981-10-0554-1_11

onward. The tokens measured 1–3 cm and were shaped in various forms, such as balls, disks, and circular cones. They retained these simple shapes until c. 4000 BC, but then came to be also decorated with lines and patterns, or shaped into the more complex forms of animals or goods. It appears that these small objects represented specific agricultural products, domestic animals, and goods and were used to indicate their number. Preceding the invention of writing in c. 3200 BC, occasionally some of these tokens were placed in a hollow clay ball, and the token shapes were impressed on the surface of the ball together with a seal impression. This set of tokens, a seal impression and a clay ball was apparently used to communicate the quantities of goods involved in a transaction and to indicate the person or institution involved in it. Similar significant finds preceding the birth of writing are the "numerical tablets" found from Uruk and Susa, which bear the impression of simple numerical signs incised by the rounded end of a stylus, together with the impression of a cylinder seal. Furthermore, some "numero-ideographic tablets" bear one or two "proto-cuneiform signs," which represent discrete objects, such as sheep, jugs of beer and dairy fat, strings of dried fruits, and products. This evidence indicates that recording devices of tokens, clay balls, seals and tablets paved the way for the invention of writing on clay tablets with the systematic use of a greater number of signs.

## 11.2 Characteristics of the Oldest Writing

The signs used for the oldest writing discovered at Uruk are pictographic and called "proto-cuneiform," since they later developed into cuneiform signs. Along with the change in the sign shapes from proto-cuneiform to cuneiform, the number of signs decreased. According to recent investigation, the oldest writing system attested in c. 3200 BC used about 1200 signs. This number had decreased to around 800 by c. 2400 BC, and eventually became less than 600. The decrease in number occurred together with the development of phonetic writing, which made it possible to describe language more precisely. The phonetic signs show that the language of the writing was Sumerian. In this development, the logographic signs, each of which represented one word, came to be used also as syllabic signs that designated phones. This may be compared to the invention of Japanese *Manyogana* and *Kana* syllabic characters on the basis of Chinese logographic characters (see below). In logosyllabic cuneiform writing, it was not necessary for every different word to be written by a distinct sign, so the total number of signs used in the writing system decreased. When this development was achieved towards the second half of the third millennium BC, cuneiform writing became more versatile and could be used for writing other languages. Thus, the system first used in Sumerian, the agglutinative language spoken in southern Mesopotamia in the fourth and third millennium BC, was then adopted in writing many other languages such as Elamite, Akkadian, Hurrian, Hittite, Luwian, and Urartian, spoken in various parts of ancient West Asia, from present-day Iran in the east to present-day Turkey and Egypt in the west. As a result, a highly developed literate culture, very old in comparison with those of other areas of the globe, was established in the broader area of West Asia. Furthermore, by the fourteenth–thirteenth centuries BC, scribes in the city state of Ugarit on the east Mediterranean coast took the linear alphabetic script used for native Semitic languages along the coast and developed an alphabetic cuneiform system with only 30 signs, using it to write their own language, and occasionally Hurrian and Akkadian. Similar phenomena occurred in the Chinese cultural sphere, where the Tangut, Khitan and Jurchen scripts were invented under the influence of the Chinese writing system.

As already mentioned, Mesopotamian writing first attested in southern Mesopotamia was invented for economic and administrative use, and the genres of documents were limited to economic and administrative texts and word lists. However, writing gradually broadened its sphere of use to include various genres, as contracts concerning the buying and selling of goods, animals and slaves, as well as those concerning marriage and loans; commercial, administrative and private letters; treaties and oaths between states, communities and persons; law codes, suits and sentences; witchcraft and anti-witchcraft literature to curse persons; incantations; diagnostic and medical texts; oracles, prophecy and omen literature; astronomical observation records; myths, epics, poems and wisdom literature; commemorative inscriptions concerning building, military and social enterprises and chronographic and historiographic works. Thus, writing eventually came to encompass almost all the areas of human activity and thought.

## 11.3 The Impact of Writing on Clay

An outstanding characteristic of cuneiform writing is its storage stability. Cuneiform script was normally written on clay tablets, which are durable against fire and do not decay. Thanks to this nature, the written sources of ancient West Asia are reported not only from exceptionally early ages, but also have survived in remarkably great numbers. Most of the documents in other ancient civilizations were written on material like papyrus, paper, bamboo cards and parchment, and have been largely lost through fire and decay. In contrast, the clay tablets of West Asia remained in the earth and have been discovered in great number from numerous excavated sites. The total number of cuneiform documents discovered so far is estimated to be as large as 500,000. This is vast in comparison with the documents of other ancient civilizations, including those of Greece, Rome and China. It is worth noting that archival texts, such as administrative-legal documents, which have rarely survived from the ancient Greek and Roman worlds, have been recovered in massive numbers from ancient Mesopotamia and Syria.

In sum, the importance of cuneiform writing is several-fold: (1) This oldest writing system influenced directly and indirectly the development of many later scripts, which eventually greatly changed human life all over the globe. (2) We can trace the process of the birth of writing from its prehistory; this is rare and instructive. (3) The documents written on clay materials have survived in great number covering many aspects of human life; thus, they supply rich information about human activities in the distant past.

All in all, the early writing of West Asia provides us with exceptionally valuable data concerning the origin and development of writing, civil life and thought, giving us a good opportunity to reflect on our modern literate world.

## 11.4 Similarity of the Cuneiform Akkadian and Japanese Writing Systems

The cuneiform writing of ancient Mesopotamia may surprise modern Japanese, since it is strikingly similar to the logosyllabic Japanese writing system, in which Chinese logographic characters (*Kanji*) and Japanese syllabic characters (*Kana*) are used in combination. It has been noticed in scholarly circles that the Japanese and cuneiform writing systems exhibit striking parallels and that their comparison provides valuable insight into the origins of early writing systems in general. For instance, in an article entitled "Bilingualism in Logographically Written Languages: Sumerian in Ebla," Miguel Civil refers to the adaptation of Chinese script in Japanese literature in the eight century A.D. He justly pointed out that "it would be useful to know what general rules prevail when languages borrow logographic scripts in order to clarify the situation in the cuneiform world" (Civil 1984:75). This is a good opportunity for us, the Japanese authors of this article, to present a short contrastive survey of the cuneiform writing system, specifically the Akkadian, and Japanese writing system.

A late Japanese linguist Rokuro Kono reached a similar view to Miguel Civil. In his contribution to *Kojiki Taisei* (Compendium of Kojiki[1]), Kono wrote: "We [Japanese] not only use two different kinds of scripts [*Kanji* and *Kana*] side by side, but also read *Kanji* in an extremely complex way not only by their *on* [Chinese(-like)] values but also by their *kun* [Japanese] values. This practice is similar to that of the Assyro-Babylonian cuneiform, which was borrowed from Sumerians. Such practice thus seems too old-looking for the second half of the twentieth century, and its complexity is unparalleled today. We struggle in this complexity day by day, but such a struggle provides us with golden opportunities for contemplating on the essence of writing." (Kono 1980)

The examples below should illustrate how similar Japanese and Akkadian writing systems are:

[1] Kojiki (古事記) or *Records of Ancient Matters* dates to 712 A.D. It preserves the mythology, the language and the traditional history of Ancient Japan. For further information, see Chamberlain's introduction to his English translation of Kojiki (1882: i–iv).

(1) Akkadian: 

| | | | | |
|---|---|---|---|---|
| Transliteration: | DUMU LUGAL | *a-na* | É.GAL-*lim* | KUR-*ud* |
| Transription: | *mār šarrim* | *ana* | *ekallim* | *ikšud* |
| Gloss: | son of king | to | palace | arrived |

| | | | |
|---|---|---|---|
| (2) Japanese: | 王子が | 宮殿に | 着いた。 |
| Transliteration: | Ō.JI-*ga* | KYŪ.DEN-*ni* | TYAKU-*i-ta* |
| Transcription: | *ōji+ga* | *kyūden+ni* | *tsui+ta* |
| Gloss: | prince+NOMINATIVE | palace+DIRECTIVE | arrive+PAST |

In example (1), the logograms are transliterated according to their Sumerian values in non-italic upper case letters, while phonograms are transliterated phonetically in italic lower case, as is customary in Assyriology. The Japanese text in (2) is deliberately transliterated in the same way, that is *Kanji*'s according to their *on* values in non-italic upper case, and *Kana*'s in italic lower case, so that anyone can recognize its similarity to the way the Akkadian text is written. Punctuation marks are used in accordance with the Assyriological convention with a single exception: the plus sign (+) in transcriptions and glosses indicates a morpheme boundary.

Both texts contain logograms and syllabograms, and these two types of letters can be mixed within a single morpheme. In these examples, we can see that the logograms represent lexemes (e.g. "son," "palace," "arrive," etc.), while phonograms spell out grammatical morphemes (e.g. the preposition *ana*, the Japanese postpositions *ga* and *ni*). Phonograms can also be used as a phonetic complement to the verb both in Akkadian and Japanese. The phonetic complements *-ud* in (1) and *-i-ta* in (2) specify some grammatical categories of the verb such as tense-aspect. Moreover, the logograms are often both polysemic and polyphonic in both systems. KUR in (1) can also be read *mātu* "land" among others. The same applies to TYAKU (着) in (2), which can be read *ki(ru)* "to wear" as well as *tsu(ku)* "to arrive."

There are two more terms that we need to elaborate. They are "***on***" and "***kun***," referred to in the above-quoted statement of Kono. As illustrated in (3) below, a Chinese character generally has two sets of readings in Japanese. *On* is the Chinese-like reading of a given letter with Japanese accent, while *kun* reads the same letter in Japanese translation. By looking at the table (3), native speakers of Chinese would recognize similarity between the *on* readings "moku" and "shin" and the Chinese pronunciation of the same letters, i.e. "mù" (木) and "shén" (神) respectively. On the other hand, the phonetic value "ki" should be totally unfamiliar to a Chinese, for it is a Japanese word meaning "tree." *kun* is the system of furnishing a Chinese character with a new phonetic value, which did not exist in Chinese, based on its semantic association with Japanese word(s). The same can be said of the cuneiform. "giš" was the original Sumerian reading, while the phonetic values *iṣ* derived from the translation of giš into Akkadian (*iṣum* "tree, wood"). This is to say that the distinction between *on* and *kun* is *not* specific to the Japanese writing system, and can be taken as a general graphemic concept.

(3) "*On*" and "*Kun*"

| *On* | Graph | *Kun* |
|---|---|---|
| moku, boku (cf. Ch. mù) | 木 | *ki* |
| giš | 𒄑 | *iṣ* |

| *On* | Graph | *Kun* |
|---|---|---|
| shin, jin (cf. Ch. shén) | 神 | *kami* |
| dingir | 𒀭 | *il* |

Nevertheless, there are three major differences between the two systems. First, Akkadian phonograms are generally polyphonic, while Japanese phonograms are not. Second, in the Japanese text, you could easily tell the difference between the logograms and the phonograms. Phonograms have relatively simple and round shapes, while logograms tend to be more complex and square shaped. Moreover, phonograms cannot be used as logograms, and vice versa. In the Akkadian writing system, on the other hand, most characters can be used both as a logogram and as a phonogram. Finally, phonetic complements are obligatory in today's Japanese orthography, while they are optional in Akkadian. However, these particular traits of the Japanese writing system did not exist in its early stage, that is, in the eighth century A.D., mainly because the exclusively phonographic *Kana* was yet to be developed at that time, and *Kanji* were employed not only as logograms but also as phonograms (*Manyogana*).[2] It follows that the similarity between the early Japanese and Akkadian systems proves to be even more striking (Ikeda 2007).

Obviously the cuneiform writing system, in which hundreds of characters are used, is not the most efficient system in comparison to the alphabetic system using merely some 30 characters. However, the world's oldest and most complex system is not exactly archaic and obsolete, as it remains instructive for linguists and all those who are interested in the variation of writing systems and their development. The old and dead cuneiform system and the living and functioning Japanese scripts lead us to reflect on the general rules for the invention, adoption and modification of the logo-syllabic writing systems that take place either among different languages or within a single language. Such reflection may further provide us with a good opportunity to ponder afresh the human activities of reading, writing, speaking and thinking.

## References

Civil, M. 1984. Bilingualism in logographically written languages: Sumerian in Ebla. In *Il Bilinguismo a Ebla: Atti del convergna Internationale, Napoli, 19–22 aprile, 1982*, ed. L. Cagni, 75–97. Naples: Istituto universitario orientale, Dipartimento di studi asiatici.

[2] *Kana* emerged out of *Manyogana* by simplifying their shapes and by eliminating polyphony and homophony.

Ikeda, J. 2007. Early Japanese and early Akkadian writing systems: A contrastive survey of 'Kunogenesis.' Paper presented at the international symposium on the origins of early writing systems, 5–7 October 2007. Beijing: Peking University.

Kono, R. 1980. Kojiki ni okeru kanji shiyō (Kanji usage in Kojiki). In: *Kono Rokuro Chosakushu* (*Collected works of Rokuro Kono*), Vol. 3, 3–53. Tokyo: Heibonsha. (Originally published in 1957)

## Further Reading

Charpin, D. 2010. *Reading and writing in Babylon*. Trans. T.M. Todd. Cambridge, MA: Harvard University Press.

Englund, R.K. 1998. Texts from the Late Uruk period. In *Mesopotamien: Späturuk-Zeit und Frühdynastische Zeit. Orbis Biblicus et Orientalis 160/1*, ed. J. Bauer, R.K. Englund, and M. Krebernik, 15–217. Freiburg/Göttingen: Universitätsverlag Freiburg/Vandenhoeck & Ruprecht.

Nissen, H.J., P. Damerow, and R.K. Englund. 1993. *Archaic bookkeeping: Early writing and techniques of economic administration in the ancient near east*. Chicago: University of Chicago Press.

Robinson, A. 1995. *The story of writing*. New York: Thames and Hudson.

Schmandt-Besserat, D. 1996. *How writing came about*. Austin: University of Texas Press.

Walker, C.B.F. 1987. *Cuneiform*. Barkley/Los Angeles/London: University of California Press/British Museum.

# Chapter 12
# Ancient Mesopotamian Gods: Mythology, Cult, and Scripture

**Uri Gabbay**

## 12.1 Mesopotamian Religion: Introduction and Sources

The study of ancient Mesopotamian religion is important first and foremost for its own sake. However, it is also important for comparative studies regarding other ancient Near Eastern religions most notably biblical religions. Since ancient Mesopotamia was the dominant political and cultural power in the ancient world until the rise of the Achaemenid and later the Hellenistic empires, it of course also exerted an influence, whether positive or negative, on other ancient societies.

We have a wealth of sources for the study of ancient Mesopotamian religion: thousands of cuneiform tablets written in Sumerian and Akkadian, stemming from about 3000 BCE up to about the first century CE, but also archaeological, non-textual evidence such as temple architecture and iconography on reliefs and cylinder seals. Despite this wealth of sources, we do not have an Akkadian or Sumerian treatise on the religious doctrines of ancient Mesopotamia that might explain their religion to outsiders such as ourselves. Rather, our knowledge of Mesopotamian religion comes from the analysis of texts written for use by the societies of ancient Mesopotamia themselves—usually, but not exclusively for religious use. For instance, Sumerian myths, which we might have expected to discover in a religious environment, such as a temple library, are actually often known to us from tablets that were written for the classroom by students who were training to be scribes; similarly, much of our knowledge about the temple cult comes not from religious texts but from administrative documents written by those who managed the large institution of the temple. Gathering all these materials is a painstaking task, but it gives us a broad view of Mesopotamian religion from multiple perspectives. Nevertheless, we must remember that we are dealing with a dead civilization, and

U. Gabbay (✉)
Department of Civilizations of the Ancient Near East, Hebrew University of Jerusalem,
Mount Scopus, Jerusalem 91905, Israel
e-mail: gabbay.uri@mail.huji.ac.il

 
A. Tsuneki et al. (eds.), *Ancient West Asian Civilization*,
DOI 10.1007/978-981-10-0554-1_12

while any treatment of religion, even modern religion, is a difficult task, it is even harder when we are dealing with the defunct religious system of ancient Mesopotamia (Oppenheim 1977: 171–83).

One thing we know for certain is that ancient Mesopotamian religion was a polytheistic religion, with thousands of gods in its pantheon.[1] Each god could have various aspects: a regional aspect—the god was related to a certain city; a natural aspect—the god was associated with an element of nature or the cosmos, such as water or the sun; and a characteristic aspect—the god was associated with a certain trait, usually of human nature or society, such as wisdom or justice. Thus, for example, the god Utu/Shamash was regionally connected to the city of Sippar, cosmically associated with the sun, and in charge of the trait of justice. The god Enki/Ea was connected to the city of Eridu, the sweet waters, and wisdom. The goddess Inanna/Ishtar was associated with the city of Uruk, the planet Venus, and love and war.

A most important feature of Mesopotamian religion, common to many religions in the world and especially the polytheistic ones, is syncretism. Syncretism refers to the identification of different gods with each other. In the previous paragraph Utu was referred to also as Shamash, Enki also as Ea, and Inanna also as Ishtar. These double names spring from a syncretism between the pantheons of the two linguistic groups who dwelled in ancient Mesopotamia in the third millennium BC: the Sumerian speakers and the Akkadian speakers. Each group had their own pantheon, but the corresponding elements of these two pantheons were closely identified with each other. Thus the sun god of the Sumerians, Utu, was identified with the sun god of the Semitic Akkadians, Shamash, and the Sumerian goddess of love and war, Inanna, was identified with the Akkadian Ishtar. Syncretism also had a local dimension. For example, the Sumerian warrior god Ninurta, son of Enlil and god of the city of Nippur, was identified with another Sumerian war god, Ningirsu, god of the city of Girsu. Similarly, various mother goddesses, Sumerian and Akkadian, were identified with each other.

The gods themselves could be related to each other as members of the same family. Often the genealogical connections between gods were derived from their regional aspects. For instance, Nabu, the god of the city Borsippa, was considered the son of Marduk, the god of the city of Babylon, since the two cities were situated in close proximity to each other.

## 12.2 The Perception of Mesopotamian Gods

Answers to questions regarding what these gods actually were and where they dwelt is not straightforward. The gods existed on several levels. First of all, let us remember that the actual place where the gods dwelt was in the minds of the people who

[1] For a recent detailed overview of the Mesopotamian pantheon, addressing some of the issues dealt with in this article, but many others as well, see Sallaberger 2004. See also Jacobsen 1970; Lambert 1975.

worshipped them, and this human perception of divinity, like many other perceptions, was multi-leveled. If we turn back to the three aspects of the divine mentioned earlier, namely, the natural, local, and characteristic aspects, we can use them to distinguish the levels on which the gods existed.

The natural aspect, for example the sun, could be the actual divine manifestation of the god associated with the sun. Thus, the god Utu/Shamash was not simply associated with the sun or in charge of it, he actually was the sun traveling across the sky. Likewise the god Enki/Ea was in fact the sweet waters, and not only the god in charge of them.

On another level, the characteristic aspects of these two gods—justice in the case of Utu/Shamash, and wisdom in the case of Enki/Ea—are reflected in myths, where the gods exist as anthropomorphic beings. The humanized gods live in a "mythical" place and time, where they act out their respective characteristics: Utu/Shamash is the divine judge, and Enki/Ea is the knowledgeable god of wisdom. This mythical place and time can also be perceived on two levels. On the one hand it is within our universe, whether in heaven, or beneath the ground, where according to Mesopotamian perception the netherworld was located, or in the sweet waters; and usually in a concrete time, in the primordial time before the creation of man or before the great flood. On another level, this existence is also beyond the earthly cosmos and outside the bounds of human timekeeping, as if it existed in a parallel dimension.

Finally, the gods have a local existence: each god or goddess is physically present in his or her locality, Utu/Shamash in the city of Sippar and Enki/Ea in the city of Eridu. The manifestation of the god's physical presence is his cult image, venerated in his temple in that locality.

Thus, each of the three aspects of divinity may be correlated with a distinct perception of a given deity: on the mythic level he (or she) is a human-like creature with a distinctive character and personality, acting to some extent beyond our natural time and space; on the cosmic level he is an element in the natural world; and on the local level he is a man-made statue or image that is the object of cultic activity.

The discussion in the following paragraphs will investigate especially the mythical and cultic aspects of the Mesopotamian divinities, dealing mainly with materials from the first millennium BC, although earlier sources will be used as well.

## 12.3 Myth: Flood and Creation of Man

Myth is usually the first element that comes to our mind when we think of ancient religions. As literary narratives that feature the gods, myths comprise the textual corpus that is most accessible and appealing to us. Actually, although myth is an important element in any religion, it may sometimes be overemphasized. For instance, myth usually does not directly dictate the religious practices of the members of a given religion, and in the case of ancient religions, the literary narratives known to us may be the product of a particular place or time, or they may be associated with one segment of society and not another, and so they are not necessarily representative of everyday religious behavior. Nevertheless, with these restrictions in mind, I would still like to begin with the mythology of ancient Mesopotamia.

We usually think of mythology in the form of written texts, but the communication of myth is not limited to a literary genre. Myth can also be represented in visual art and in other forms, although it is true that it received its fullest expression in the genre of epic narrative. In this article I will define myth as a narrative that features a characteristic trait of a god. This trait is often exaggerated in the narrative. Thus, if a god is perceived to be capable of causing terrible catastrophes, including natural disasters such as floods, a myth about the god may narrate the story of a specific great flood—*the* flood that the god brought upon the earth.

The example of the flood was not chosen at random. I would like to focus on the story of the flood, since this is one of the most well-known myths of ancient Mesopotamia, and has some striking parallels to the biblical flood narrative (Finkel 2014). The myth appears in its best known form in the standard version of the Gilgamesh epic, which dates from the first millennium BCE (George 2003). In this epic, the legendary Gilgamesh, king of the city of Uruk, sought eternal life. During his quest he reached Uta-napishtim, the only survivor of the ancient great flood. In the eleventh tablet of the epic, this Uta-napishtim tells Gilgamesh about the flood, how it killed all of humanity except for him and his family, and how he built a ship to preserve himself, his family, and a selection of animals, and eventually received eternal life. This myth has of course found its way into the Bible through its story of the flood and its protagonist Noah.

The story Uta-napishtim tells Gilgamesh is heavily dependent on another, older myth, known to us from tablets from the early second millennium BCE (Lambert and Millard 1969). This myth, called the Atra-hasis myth, begins with the creation of mankind and ends with the devastating flood that the gods unleashed on the humans they themselves had created. The myth begins in primordial times, when there were no humans, only gods. In those days, the myth says, the gods had to toil very hard. They had to toil hard due to the natural and ecological environment of Mesopotamia, where irrigation and agriculture mainly depend on manpower. Abundant water is provided by the Tigris and Euphrates rivers, but unlike the rainwater that falls from the sky, the water in the rivers has to be brought to the fields by digging canals. So the gods had to do this work before man was created. This is because the gods, like man, had to eat and drink, and the only way in Mesopotamia in which human beings can have a steady supply of food and drink is through physical toil, especially canal digging; and gods, like humans, have to eat as well, and therefore they had to work for their food. So, the myth continues, the gods were working hard, but they were tired, and the greater gods decided to create someone who would do their work for them: man. Man was created by the god Enki/Ea and the Birth Goddess, and he took over the work of digging canals. However, although the gods had freed themselves from hard labor, they were not freed from the need to eat. So it was man's responsibility to feed the gods which was done through offerings placed before the physical images of the gods in their temples (Maul 2008). Thus man is not a free creature, for he must work not only to feed himself, but to feed the gods as well. The gods, however, are not free either, since they are dependent on man for their food supply.

This perception of the gods as dependent on humans is developed later in the myth, in the flood episode. After the great gods create man to do their work, the human population steadily increases. This increase in persons leads to more and more earthly noise, which disturbs the sleep of the great gods. The gods thus decide to destroy these noisy creatures. They make several attempts to destroy them with plagues or famine, but ultimately they decide to destroy man with a great flood. After the gods swear to bring the flood, the wise god Enki/Ea decides to warn the sage Atra-hasis, who is identical with Uta-napishtim of the Gilgamesh epic, about the coming catastrophe. Enki/Ea commands Atra-hasis/Uta-napishtim to build a large boat to save himself, his family, and a selection of animals. The devastating flood destroys all the land's inhabitants, except for the ship and its passengers. When the flood ends and the waters have receded, Atra-hasis/Uta-napishtim exits the ship and presents an animal offering to the gods. Here we see the consequences of the gods' dependence on man for food. The gods, during the entire period of the flood, could not eat. There were no men to provide offerings, and they became very hungry. Therefore, when Atra-hasis/Uta-napishtim makes his offering, the myth describes the scene thus (Lambert and Millard 1969: 98–99; George 2003: 712–13):

> The gods sniffed the smell,
> They gathered like flies over the offering.

The imagery used here dramatically portrays the gods as hungry beings, descending like flies on the meat. This story, then, not only emphasizes the devastating nature of the gods, who can bring a flood upon mankind; it also indicates the crucial function of the daily offerings to the gods, provided by man in the gods' earthly temples.

## 12.4 Cult: The Divine Image

I would now like to turn from mythology to focus on the Mesopotamian temple cult. First, a word about ancient Mesopotamian temples: One of the most well-known features of Mesopotamian temples is their ziggurat towers. However these towers were not the place where the regular cult occurred. The regular cult occurred in smaller structures, which often stood beside the ziggurat. These structures were modeled on the architecture of a regular house, since the temples were first of all the earthly dwellings of the gods, where they ate, drank, and slept. As we have seen in the mythological account of the flood, the personified gods were dependent on earthly food, the offerings presented to the gods in their temples. In the daily cult, the god received these offerings through his cultic image. This image, usually anthropomorphic and made from wood and stone, sat in the innermost cella of the temple, where the priests presented offerings and recited prayers before it. It is important to stress that the priests did not regard the statue as a symbol of the divinity's invisible presence in the temple—they recognized it as the divinity himself, manifesting himself in one of his aspects.

This cult directed to a statue was the target of much mockery in the Bible which questioned how could a statue eat or hear, even if it has ears and a mouth. This is explicitly noted in the book of Psalms (115: 4–7)[2]:

> Their idols are silver and gold, the work of men's hands.
> They have mouths, but they speak not; eyes have they, but they see not;
> They have ears, but they hear not; noses have they, but they smell not;
> They have hands, but they handle not; feet have they, but they walk not; neither speak they with their throat.

Indeed, the theologians of ancient Mesopotamians had to deal with this religious issue. It is debatable whether they were so naive as to think that a statue made by man could actually hear their prayers and eat their offerings. Again, we have no theological treatise from ancient Mesopotamia that might explain how they dealt with this problem, but we can get some insight into their thought by examining the ritual in which the divine image, made by human hands in a workshop, was transformed into a god worthy of cult. This ritual, lasting two days, was called the "mouth washing" ritual and included seven washings of the mouth of the image. I will summarize the ritual according to tablets found at Nineveh that were written mainly in the seventh century BCE (Walker and Dick 2001: 33–67)[3]:

On the morning of the first day, before dawn, the incantation priest, the priest who will be in charge of the entire ritual, goes out to the riverbank and marks the exact spot where the sun rises with a stone. The reason for this will be evident below. He makes an offering at that spot and draws water from the river into seven bowls. He fills these bowls with plants, herbs, various stones, oils, honey, and butter, and wraps the bowls with red, white, and blue wool. He waves incense over the bowls while reciting incantations and offerings, and then returns to the city with the bowls.

Back in the city, the priest enters the workshop where the statue of the god was manufactured and performs the first mouth washing on the statue, using the first bowl brought from the river. He says to the cultic image (Walker and Dick 2001: 57):

> From this day you will go before your father, the god Ea. May your heart be pleased, may your mind be happy! May your father Ea be full of joy with you.

The priest takes the statue out to the river, reciting incantations on the way. The entire ritual is accompanied by many incantations, prayers, and offerings that are of great religious importance. The statue has already gone through the first mouth washing and is therefore not just a chunk of wood any more. On the other hand, it is not yet fully divine either, a state that will be achieved only after the seven mouth

[2] For biblical polemics against the cultic image, see Berlejung 1998: 315–413; Dick 1999.

[3] For various studies of this ritual, see Berlejung 1997, 1998; Walker and Dick 1999, 2001. The following analysis of the ritual is partially based on the discussions in these publications.

washings have been completed. Furthermore, the divine image of the god has a specific place in the human world: in the temple, where he receives cult. At this point in the ritual we have a statue that is only partly divine, and is not yet in its natural place, the temple: it is about to be carried from the workshop to the river. This is a hazardous situation, since in religion, as in all human society and behavior, man prefers "either/or" situations—in this case: either the statue is a god, or it is not a god—and not the grey areas in between. These grey areas are often perceived by humans as threatening, and therefore in order to protect humanity, and specifically to protect the priest and the statue in this hazardous situation, prayers and incantations are recited, and offerings are presented throughout the ritual.

The priest and the statue do not proceed alone to the river. The craftsmen who built the statue in the workshop accompany them to the river in a cultic procession. When they arrive, the priest sacrifices a ram. He then takes the thigh of the ram and inserts the craftsmen's tools—an ax, a chisel, and a saw—into the thigh and throws it into the water. The sweet waters of the river are the domain of the god Ea, the god of wisdom and technical skill, who oversees the creation of the statue. The act of throwing the tools into the river signifies that the tools used by the craftsmen to construct the image are considered divine, the property of the god Ea. Once the statue has been completed, the tools are returned to the god Ea together with an offering. Thus, this ritual act attempts to deal with the paradoxical creation of the god by human craftsmen by asserting that the tools used in his creation belong to the divine world, specifically to the god Ea.

Once the tools have been properly disposed of, the priest performs the second mouth washing and recites the following incantation to the god Ea three times (Walker and Dick 2001: 59):

> He who comes, his mouth is washed! May he be counted among the gods, his brothers! Take away the ax, chisel, and saw of the craftsmen, all that came near him, from its (i.e., the ram's) body! This god, oh Ea, is washed! Count him with his brothers!

This prayer expresses the wish that the statue will become a god like his fellow gods. The priest now seats the statue on a reed mat in a garden near the river, directing the statue's eyes towards the east, to the exact spot where the sun will rise, according to the stone marker the priest placed on the riverbank early in the morning. Here the priest performs the third and fourth mouth washings, and then the statue remains on the riverbank all night until dawn, when he watches the sun (or better: the Sun-god) rise, a symbol of renewal and divine manifestation, manifesting his divine birth with the new day.

However, the statue is not yet a complete god: he cannot hear, and he cannot eat. However, there are still three mouth washings left. In the morning the priest performs offerings and recites an incantation to three deities—Ea, the sun god Shamash, and the god of magic Asarluhi—from which a few lines will be cited (Walker and Dick 2001: 134, 149):

> … (Oh gods), be present on this day! Be present for this statue before you!
> Determine its destiny greatly that its mouth may eat,
> That its ear may be set for hearing! …

The great gods are asked to "determine the destiny" of this statue so that it can eat and drink and hear—or, in other words, so that the statue can hear the prayers directed to him and eat the offerings presented to him.

Later, the priest performs the fifth mouth washing and recites another incantation, from which I will cite a few passages (Walker and Dick 2001: 135–39, 149–50):

> When the god was fashioned, the pure statue was completed,
> The god appeared in all the lands. …
> It was created in heaven; it was created on earth.
> This statue was created in the entire heaven and earth …
> The statue is an image of the gods and of man …

Again we see here an attempt to grapple with the theological paradox that the statue was made on earth but is nevertheless divine. According to this incantation the statue was made both in heaven and on earth that is, by both the gods and man, and accordingly has the features of both. The incantation proceeds (Walker and Dick 2001: 140–41, 151):

> … This statue cannot smell incense without 'opening the mouth',
> It cannot eat, cannot drink water.

Here, too, the fact that the statue is not yet fully animate and cannot eat, drink, or smell is acknowledged, almost in the same phrases used in the biblical polemic against this belief. However, the recitation of this incantation, along with the other cultic acts, will give the statue these powers. In just a few hours the statue will be able to smell incense and eat and drink, and, as the direct manifestation of the god, to participate in cult.

The priest now performs the sixth mouth washing and whispers into the right ear of the statue (Walker and Dick 2001: 65):

> You are now counted among the gods, your brothers!

The ritual is not over yet. Now a dramatic act occurs. The priest places the craftsmen who created the statue in front of altars directed to the gods in charge of the arts practiced by the craftsmen. The priest binds the hands of these craftsmen and then cuts them off—symbolically— using a wooden sword. At the same time, the craftsmen say the following while their hands are supposedly being cut off (Walker and Dick 2001: 66):

> I did not create the statue! I did not create the statue at all! It was [the craftsman god] who created the statue![4]

As in the earlier symbolic action, when the tools of the craftsmen were thrown into the river and returned to the god Ea, here too the cutting off of the hands symbolizes that the creation of the god was not a human creation, but a divine creation. The tools thrown into the river could be replaced, but to actually cut off the hands of the

[4] This translation is a paraphrase. The original text names the various gods that are associated with the different craftsmen; see Walker and Dick 2001: 66.

talented craftsmen who manufactured the statue would not be sensible, either pragmatically or economically. Therefore this is a symbolic act, but the violence it represents indicates just how important it was to deny any connection between the human hands of the craftsmen and the divine creation of the gods.

The priest now heads back to the city. Upon entering the temple, the priest recites the following incantation (Walker and Dick 2001: 177–79, 187):

> My lord, the temple was built for you for your happiness!
> … When you enter into your temple –
> May your heart be pacified, may your mind be calm!
> May a good fate be determined for you!
> Pass the days in the temple in gladness!
> May the god Anu provide you food rations!
> May the god Enlil present to you offerings and flour!
> May a good spirit stand before you,
> And may you hear its prayer!

This last plea to the god emphasizes once again his cultic role: the god should happily dwell and rest in his temple, eat the food that has been allotted to him by the main gods and that is delivered by the humans who present him with offerings, and listen to the prayers recited by the good spirits who represent humanity.

At last the priest seats the god on his throne in the temple, recites the final incantations, and makes the last offerings. He performs the seventh and final mouth washing and dresses the god in his crown and garments. The image of the god is now finally a complete god worthy of cult.

We have seen, then, that although it is paradoxical to imagine a statue eating offerings and hearing prayers, this was a crucial part of the way in which divinity was perceived in ancient Mesopotamia. The long ritual described above was meant to cope with this paradox, and to give the god the powers he needed to exist not only in the mythical world, but in the earthly, human, and physical world as well.

## 12.5 Scripture: Gods and Texts

Lastly, there is one more aspect of the divine that I would like to address, although very briefly. We have already seen the god Enki/Ea involved in two creations: the creation of man in the introduction to the Atra-hasis flood myth, as well as the creation of the statue of the god in the "washing of the mouth" ceremony. However, there is another creation that Enki/Ea was in charge of. The god Ea was the god of wisdom, and wisdom, in the literate society of Mesopotamia, was closely associated with the written word and with texts. The most important texts of ancient Mesopotamia, including liturgy and incantations, as well as divinatory texts such as astrology, and some myths too, were considered to be creations of the god Ea (Lambert 1962, 62). The texts that were attributed to the god Ea comprised the main part of the Mesopotamian textual canon in the first millennium BCE, and the attribution of these texts to a divine composer, the god Ea, confirmed their importance and authority. The beginning of this article dealt with three main aspects of the divine: the natural, the mythical, and the local concrete-cultic. However, with the process of textualization that Mesopotamian society, and especially its accumulated knowledge, underwent, another aspect of the divine grew important too: the textual or scriptural aspect.

Today, the association of wisdom and knowledge with textuality seems natural to us but there is no necessary connection between the two. Wisdom can also be transmitted orally, and when it is transmitted in writing, there is no reason to think that the text is anything more than a simple medium of transmission. In Mesopotamia, where written texts were a defining part of its culture from such an early date, textuality gained religious importance as well. Clay tablets were not simply a medium through which knowledge was transmitted. They provided authoritative textual knowledge that was often perceived as coming from the gods themselves.

The association of textual knowledge with divine authority has many implications: It discourages humans from altering texts, and therefore is closely connected to the process of canonization. In addition, the perception that an inscribed text carries the authoritative divine word is closely connected to the process of interpretation and hermeneutics that seeks to explain and decode the written divine message, and indeed Akkadian commentaries on the canonical divine texts would develop during the first millennium BCE (Frahm 2011).

Finally, the perception that gods, like human beings, communicate through written texts is reflected in mythology. In many myths, the divine authority can only be carried out by one who possesses an object called "the tablet of destinies." This is clearly seen in the Epic of Anzu, known from the first half of the second millennium BCE onwards. This epic tells about the battle between the war-god Ninurta and the mythological thunderbird Anzu (Annus 2001). The reason for this battle is that Anzu stole the tablet of destinies from the head of the pantheon, the god Enlil, and thus acquired power for himself. Ninurta, Enlil's son, is sent to battle against Anzu in order to recover the tablet of destinies and return it to his father Enlil, a mission that he indeed successfully accomplishes. A similar scenario, but featuring the god Marduk, is also found in the Babylonian creation myth Enuma elish, known from many sources from the first millennium BCE (Lambert 2013: 451).

In Assyrian religion, which was very much connected to royal ideology, the perception of the tablet of destinies went one step further. The divine tablet of destinies had an actual physical manifestation as well. Just as the gods were perceived both mythically and physically, as discussed above, so also the tablet of destinies was not only a mythical tablet existing in the realm of gods, but had a physical existence too. The treaties by which vassal countries subjected themselves to Assyrian imperial rule—written on actual tablets that spelled out the obligations of these vassals—were also considered divine "tablets of destinies": they were sealed with a seal that, according to the inscription on it, was owned by the god Ashur himself, and they were displayed in temples (George 1986). Thus, the political obligations of human beings were considered to be among the "destinies" written down on the divine "tablet of destinies."[5]

Perhaps, then, Mesopotamian religion was actually the first religion of the book, or better, of the clay tablet.

## References

Annus, A. 2001. *The standard Babylonian Epic of Anzu, State archives of Assyria texts, vol. 3*. Helsinki: Neo-Assyrian Text Corpus Project, University of Helsinki.

Berlejung, A. 1997. Washing the mouth: The consecration of divine images in Mesopotamia. In *The image and the book: Iconic cults, aniconism, and the rise of book religion in Israel and the ancient near east*, Contributions to biblical exegesis and theology, vol. 21, ed. K. van der Toorn, 45–72. Leuven: Peeters.

Berlejung, A. 1998. *Die Theologie der Bilder: Herstellung und Einweihung von Kultbildern in Mesopotamien und die alttestamentliche Bilderpolemik*, Orbis Biblicus et Orientalis, vol. 162. Freiburg/Göttingen: Universitätsverlag/Vandenhoeck-Ruprecht.

Dick, M.B. 1999. Prophetic Parodies of making the cult image. In *Born in Heaven, made on Earth: The making of the cult image in the ancient near east*, ed. M.B. Dick, 1–53. Winona Lake: Eisenbrauns.

Finkel, I. 2014. *The Ark before Noah: Decoding the story of the flood*. London: Hodder and Stoughton.

Frahm, E. 2011. *Babylonian and Assyrian text commentaries: Origins of interpretation*, Guides to the Mesopotamian textual record, vol. 5. Münster: Ugarit-Verlag.

George, A.R. 1986. Sennacherib and the tablet of destinies. *Iraq* 48: 133–146.

George, A.R. 2003. *The Babylonian Gilgamesh epic: Introduction, critical edition and cuneiform texts*. Oxford: Oxford University Press.

Jacobsen, Th. 1970. Mesopotamian gods and pantheons. In *Toward the image of Tammuz and other essays on Mesopotamian history and culture*, Harvard Semitic series, vol. 21, ed. Th. Jacobsen and W.L. Moran, 16–38. Cambridge, MA: Harvard University Press.

Lambert, W.G. 1962. A catalogue of texts and authors. *Journal of Cuneiform Studies* 16: 59–77.

Lambert, W.G. 1975. The historical development of the Mesopotamian Pantheon: A study in sophisticated polytheism. In *Unity and diversity: Essays in the history, religion, and literature of the ancient near east*, ed. H. Goedicke and J.J.M. Roberts, 191–200. Baltimore: Johns Hopkins University Press.

Lambert, W.G. 2013. *Babylonian creation myths*, Mesopotamian civilizations, vol. 16. Winona Lake: Eisenbrauns.

Lambert, W.G., and A.R. Millard. 1969. *Atra-Ḫasīs: The Babylonian story of the flood*. Oxford: Oxford University Press.

Maul, S.M. 2008. Dem Gott ernähern. Überlegungen zum regelmässigen Opfer in altorientalischen Tempeln. In *Transformations in Sacrificial Practices from Antiquity to Modern Times: Proceedings of the International Colloquium, Heidelberg, 12–14, Juy 2006*, ed. E. Stavrianopoulou, A. Michaels, and C. Ambos, 75–86. Berlin: Lit Verlag.

Oppenheim, A.L. 1977. *Ancient Mesopotamia: Portrait of a dead civilization* (revised edition). Chicago: University of Chicago Press.

Sallaberger, W. 2004. Pantheon. A.I. In Mesopotamien. *Reallexikon der Assyriologie* 10: 294–308.

[5]For the political and religious implications of the Assyrian "tablet of destinies," see Watanabe 2014: 157–65.

Walker, Ch., and M.B. Dick. 1999. The induction of the cult image in ancient Mesopotamia: The Mesopotamian *mīs pî* ritual. In: *Born in Heaven, Made on Earth: The making of the cult image in the ancient near east*, ed. M.B. Dick, 55–121. Winona Lake: Eisenbrauns.

Walker, Ch, and M.B. Dick. 2001. *The induction of the cult image in ancient Mesopotamia: The Mesopotamian Mīs Pî ritual*, State archives of Assyria literary texts, vol. 1. Helsinki: The Neo-Assyrian Text Corpus Project. University of Helsinki.

Watanabe, K. 2014. Esarhaddon's succession oath documents reconsidered in light of the Tayinat version. *Orient* 19: 145–170.

## *Further Reading*

Mesopotamian gods and religion: Jacobsen 1970; Th. Jacobsen, *The treasure of darkness: a history of Mesopotamian religion*, New Haven: Yale University Press, 1976; J. Black and A. Green, *Gods, demons and symbols of Ancient Mesopotamia: an illustrated dictionary*, Austin: University of Texas Press, 1992. The Babylonian flood myth: Finkel 2014. English translations of Mesopotamian myths and legends: B. Foster, *Before the Muses: an anthology of Akkadian literature*, 3rd ed., Bethesda: CDL Press, 2005. The induction of the ancient Mesopotamian cult image: Walker and Dick 1999.

# Part IV
# Importance of Cultural Heritage

# Chapter 13
# Do Archaeological and Conservation Sciences Save Cultural Heritage?: Cultural Identity and Reviving Values After Demolishment

**Yoko Taniguchi**

## 13.1 Introduction

The conservation of cultural heritage is most challenging. In areas like West Asia where prolonged internal conflict and attacks on tremendously valuable cultural heritage sites are ongoing, conservation may even seem hopeless at times. Therefore my immediate and honest answer to the question of this paper's title is 'no'. Basically, conservation science is a field that aims to study archaeological, artistic or architectural cultural materials based on the use of scientific methods for identifying creative techniques and causes of deterioration. It also aims to distinguish suitable methods and materials for conserving cultural heritage through study of the causes of deterioration, measures for improving materials, and techniques for conservation.

However, the harsh reality is that no conservation science has yet been able to pinpoint measures for ending or curtailing the destruction of cultural heritage and looting of archaeological sites. Notably, extremist organizations began indiscriminately targeting cultural materials for demolishment in 2003, after the Iraq war. These violent assaults and conditions of unrest are fuelled by highly complex concerns touching on religious, ethnic, political, economic, and other matters that—owing to a global field of growing injustice and inequality—remain too difficult to solve.

Once a conflict or a war occurs, archaeological sites, historical monuments and museum collections suffer, often as direct victims, even where demolition is enacted as a form of symbolic *propaganda* to send a message to both local and international parties, as was witnessed in Bamiyan, Afghanistan. Well planned videos by ISIL of the destruction of the Mosul museum, Nineve, and Nimurd in Iraq in early 2014,

Y. Taniguchi (✉)
Faculty of Humanities and Social Sciences, University of Tsukuba,
1-1-1, Tennodai, Tsukuba 305-8571, Ibaraki, Japan
e-mail: taniguchi.yoko.fu@u.tsukuba.ac.jp



A. Tsuneki et al. (eds.), *Ancient West Asian Civilization*,
DOI 10.1007/978-981-10-0554-1_13

along with images of a demolished Palmyra in Syria, were successful in generating a strong emotional response—and sense of futility—globally. Such destruction is essentially an example of 'performative acts of producing imagery of violence in the public sphere' whereby each attack on a cultural image—framed in the logic of iconoclasm—serves as a discursive tool (Harmanşah 2015). These acts are referred to as '*iconoclash*' (Latour 2002).

Equally, it is important to recognize that U.S. and allied military forces in Iraq also bear responsibility for the destruction and looting of archaeological sites and objects (Rothfield 2009: 3–9; Gibson 2008: 38–39). For instance, in 2003, U.S. and allied Forces deliberately damaged Babylon as they built the Camp Alpha military base (Bahrani 2013). Official criticism of this or other transgressions by international organizations such as UNESCO did not emerge until 2009. In addition, illicit excavations and lootings are increasingly driven by growth in the global commercial art market. No matter how one views it, the presence of conflict does not give rise to *any* positive outcomes for cultural materials.

> "It is easy enough to view the damage to these heritage sites as something that can be rebuilt, if all we see is the physical; it is far more difficult to grasp the effects of cultural cleansing in the longer term, and to truly absorb its impact on a country's people" – Lynda Albertson, Chief executive of the Association for Research into Crimes against Act (Shaheen 2015).

## 13.2 Destruction as *Cultural Cleansing* and *Cultural Genocide*

There are many well-documented cases of radical action taken to destroy icons imbued with collective cultural meaning from history, as occurred in Bosnia and Herzegovina in 1992–1995 (Haramoto 2009). In this case, material items and physical locations that played foundational roles in local cultural identities, such as cemeteries, religious buildings, libraries, and symbolic monuments, were targeted for destruction. Such cultural institutions generally stand as instruments for accessing the memory of previous generations. Of course, archaeological and historical artefacts are also often at the root of people's cultural identity, even where religious belief or ethnic composition is no longer identical to that prevalent at the time artefacts were created. Therefore, cultural materials are vulnerable to exploitation as apparatuses for delimiting the boundary between generations and ethnicities.

The threat of cultural genocide (Brent 1994a, b) looms, manifest in the scattering of cultural materials through a vast international art market comprised in part by individual buyers. This dispersion is not limited to material objects alone, such as at archaeological sites and monuments, but also includes all types of intangible traditions and practices, as well as the continuity of common memories and education.

At the same time, museums in conflict zones are often faced with the threat of inventory loss, not only through pillaging and destruction of museum collections, but also through damage and disorder rendered to records, leading to the loss of

**Fig. 13.1** After looting at the Kabul National Museum in Afghanistan (Photograph: Afghanistan National Museum)

information of item origins and contexts (Fig. 13.1). The Kabul National Museum (currently the National Museum of Afghanistan) lost its inventory records during the civil war of the 1990s, and important information for a substantial portion of its objects is still missing today. Conservation terminology tends to commonly reference actual interventions on material objects, such as the re-jointing of broken fragments. However, less drastic action is usually far more effective for the purpose of museums and archaeological sites: namely, the utilization of inventory, archiving and documentation (Cather 2010). The Society for the Preservation of Afghan Cultural Heritage (SPACH), the Oriental Institute of Chicago University for the National Museum of Afghanistan, UNESCO, the German Archaeological Institute, and the Museum of Islamic Art in Berlin, among other international organisations, universities and institutes, have been cooperating with damaged museums in Iraq, Syria and Afghanistan to create new databases and inventories.

By the initiative of UNESCO, the 1954 Hague Convention for the Protection of Cultural Property in the Event of Armed Conflict and the 1999 Second Hague Protocol were issued in order to protect looted objects, including prohibition of importation from countries under conflict. A special emblem, the so-called 'blue-shield', was also selected to identify cultural properties in need of safeguarding. However, the Convention and the Protocol have not been fully operational with regard to recent conflicts, especially those in West Asia, as they only technically apply to instances where war exists between nations and not instances of internal conflict and strife. In reality, neither international frameworks nor conservation efforts function to capably restrain demolishment and looting of sites and museums.

## 13.3 After the Destruction of Afghanistan's Giant Buddhas

Here I would like to use an example from Afghanistan to illustrate a case where conservation and archaeological science have played a role in rehabilitating a site following destruction. From the time of the 1978 Saur Revolution and invasion by the USSR, Afghanistan grappled with over 10 years of conflict and civil war. Many Afghanis, including the intellectual class, left the country and its interior regions fell into disorder. Following intervention by US-led Allied Forces and the Northern Alliance after September 11, 2001, Afghanistan has experienced exacerbated insecurity marked by numerous small conflicts between the ISAF (International Security Assistance Force) and the Taliban Emirate. Since 2011, security levels once again fell as the Taliban and other anti-government groups grew in power.

The destruction of the Great Buddha statues by fundamentalists has proven that such anti-cultural activities are indeed effective in capturing attention and spreading their message to the world. These acts were not simply iconoclastic responses to non-Islamic 'idols' belonging to Buddhism, but were also strongrebukes against international bodies concerned with cultural property above other crucial political issues in Afghanistan. Also, they arguably served the purpose of strategically representing the Taliban's power in the historically contested Hazara region of Bamiyan. Thus, this particular case should be understood as a means to spread organizational propaganda and is thereby distinguishable from the cultural genocides that have wracked Bosnia and Herzegovina, as well as contemporary Syria, Iraq and others (Fig. 13.2).

Historically, the importance of Bamiyan's cultural legacy is manifest in the two colossal carved Buddha images at the eastern and western ends of a high cliff made of friable conglomerate rocks facing the central valley. There are also hundreds of caves cut into the cliff face, of which about 50 contain a rich variety of wall paintings. The Buddhist tradition of Bamiyan, dating mainly between the fifth and ninth centuries ACE, was part of a renaissance in Central Afghanistan following the collapse of the earlier Gandharan culture, which crossed numerous borders as it spread through and influenced various countries along the Silk Road. Afghanistan itself can be characterized as a missing link between civilizations of the East and West—as represented in the Buddhist culture of India and the culture of Greece, respectively and from which their confluence then spread on towards China (Miyaji 2002:42).

**Fig. 13.2** Bamiyan cliff and empty Buddha niches from Shar-e-Gholghola (2013)

While Bamiyan is a medieval Buddhist archaeological site, there are no known Buddhists in the region and the area is mainly populated with Twelver Shia Muslims. (Local inhabitants known as the Hazaras are, however, Mongoloid.) Although no groups have engaged in worship of the Giant Buddha statues, they had nevertheless been taken care of by locals over the centuries. Locals often gathered in front of the cliff for the spring festival (Nauroze), football games, and on other occasions. The cliff with the Buddhas was visible from anywhere in Bamiyan. The statues and wall paintings did not function as religious symbols, but were a familiar aspect of the landscape of Bamiyan for centuries. They also provided significant income to locals as a major tourism resource.

Over the course of Afghanistan's civil war, apart from the two Giant Buddhas, countless smaller Buddhas in Bamiyan and Kakrak valley were also destroyed, while wall paintings and reliefs in the caves were similarly subject to cutting and looting. The Kabul National Museum was used as a military base in the early 1990s, at which time many precious objects were looted and sold to overseas art markets including those in Tokyo (JCICC 2004).

One of the most famous collections from the museum—the Bactrian treasures of TillyaTepe, comprised of extraordinary gold, silver, ivory and turquoise ornaments—were, on the other hand, carefully hidden in 1988 by order of President Mohammad Najibullah. They were secretly kept in 6 sealed safes located at the Presidential Palace and the Ministry of Information and Culture. These archaeological treasures were held in good condition for about 15 years during the war without incurring damage or being lost. Such is one rare case of the successful protection of cultural materials by government and local academics in a period of conflict and confusion.

Despite the loss of the Buddhas in 2001, as well as the vandalism and looting that followed the explosions, the remaining niches of the colossal Buddhas and the caves around them still stand as a testimony to the important historical value of this Buddhist site. The Bamiyan site embodies more than just a marker of Buddhist culture that is no longer locally practiced as a religion. It is important both historically and artistically for its place along the ancient routes of cultural and commercial exchange on the Silk Road linking the East and the West. The site was inscribed on the World Heritage List in Danger in 2003 and thus designated as subject to protection under international cooperation. A conservation project, titled 'Safeguarding of the Bamiyan Site' and executed within the framework of UNESCO, was organised for the sake of assuring protection through international cooperation. The project, funded by the UNESCO/Japanese Funds-in-Trust and the National Research Institute for Cultural Properties, Tokyo (NRICPT) incorporated the involvement of other institutions and universities, such as the International Council on Monuments and Sites (ICOMOS) Germany, Aachen University, and Italian institutions, from its early stages (JCICC 2004).

Since 2004, as a part of the project, a team from ICOMOS Germany has been working on securing damaged Buddha niches that still contain remains of the Buddha statues (Petzet (ed.) 2009). All statue fragments have been inventoried and removed from the niches. The niches were then consolidated geomechanically with well-planned anchoring and grout injection. Over the course of the work, a number of painted surface fragments from the Buddha statues were found and saved along with fragments from wall paintings located on niche ceilings. In fact, explosion blasts were found to have completely shattered valuable wall paintings in niches situated above the statues. One of Bamiyan's most important paintings, a 'decorated Buddha', once existed above the East Giant Buddha but no longer survives. In addition to paintings in the niches, other valuable paintings in the caves have also been vandalized or stolen for sale in overseas art markets including those in Japan, the UK and the US. Important figures have been cut along lines in the images using knives and chisels. Numerous bullets and bullet holes are still visible in the paintings. It is believed that these acts largely took place during Afghanistan's prolonged decades long civil war. On 5 August 2015, Japan announced that it would return to Afghanistan 42 pieces of wall paintings from Bamiyan and Foladi caves (102 pieces in total including items from the Kabul National Museum) that had been found in Japan.

## 13.4 Reviving Cultural and Scientific Values After Demolishment of the Bamiyan Site

Recently, the two Giant Buddha statues and over 50 caves from the Bamiyan site were scientifically dated using the $^{14}C$ technique at Nagoya University (Yamauchi ed. 2006). Spurred by extended debate around dating in archaeology and art history (Miyaji 2006; Iwai 2007), utilization of this scientific approach during conservation provided significant opportunity to develop a clear chronology of ancient Bamiyan.

Shortly thereafter, the Italian company TREVI successfully executed a geo-engineering consolidation of the damaged main cliff between 2003 and 2006 (Margottini 2009; Crippa et al. 2013). Aachen University in Germany and the NRICPT later initiated the creation of a Cultural Masterplan, although it met with challenges stemming from rapid urban development such as the unauthorised construction of hotels within the buffer zone.

The intact portions of wall paintings in the caves have suffered extensively from deterioration and destruction owing to their inherent material frailty, extreme environmental conditions and, above all, human encroachments. Most of the paintings have become almost entirely detached from the friable conglomerate rock fabric and are covered with black soot-like deposits from past human activity. It is believed that this is the most characteristic aspect of the Bamiyan site (Taniguchi and Aoki 2005). Difficulties faced in the conservation of mural paintings in Bamiyan arise in not only physical and chemical considerations, but also technical, ethnic and political matters that are not easily resolved (Taniguchi 2007).

Urgent protection for damaged Buddha statues and wall paintings has been essential to protect them from further destruction or looting. Therefore, securing fragments of both statues and wall paintings has been prioritised, with fragments transferred to safe storages (JCICC (ed.) 2007a, b). The inventory of fragments covers over 10,000 pieces and condition surveys that list the degree of deterioration as base data (JCICC ed. 2012; Yamauchi ed. 2013).

ICOMOS Germany provided identical samples from the surface of two Buddha statues for analysis, different from areas previously restored by the Archaeological Survey of India (ASI) in the 1960–1970s. These were studied in Japan and the United States using PLM, SEM-EDS, GC/MS and nano-LC-MS/MS techniques. Results were of an interesting nature and showed that the East Giant Buddha had been painted multiple times in red, blue, and yellow on a white gypsum ground with water soluble binder, containing xylose and amino-acids (Taniguchi 2010: 99–100). The origin of the amino-acids, either from egg or other animal/fish-based glue, still remains unidentified at this time. On the other hand, the West Giant Buddha was painted once with red mixed with cow glue (Fig. 13.3). It was previously believed that cow based animal hide glue could not be used in Buddhist contexts such as Bamiyan. Thus, results gave new insight into the pallet of ancient artisans employed by the kings and monks of ancient Bamiyan.

These analyses also connect numerous ancient sites along the Silk Road. In particular, the use of highly sophisticated painting techniques—such as multi-layered painting structures allowing for the creation of various colour hues and tones, lead white in drying oils, and copper salts for greens—are our very first links in Central Asia linking the painting technologies of East and West (Figs. 13.4 and 13.5) (Taniguchi 2012a, b).

Owing to damage at the Bamiyan site, much of the Buddha statues and wall paintings have been lost; however, even subsequent to destruction, extensive information on constituent inorganic and organic materials, technologies of creation, original colours, and chronological data with $^{14}C$ has been revealed. The present may just be the most well studied period of the site since it was created.

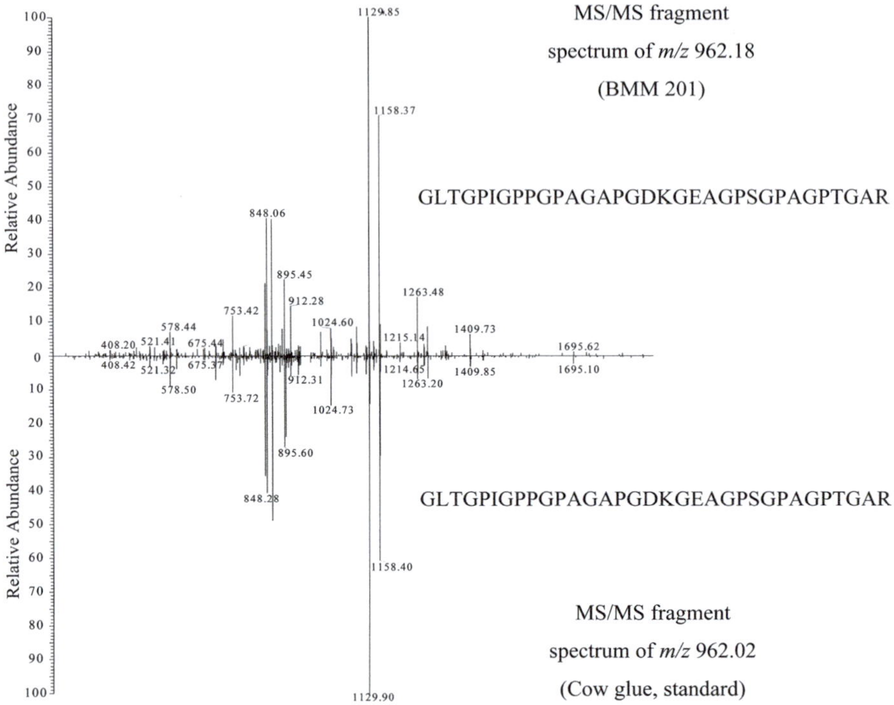


**Fig. 13.3** Comparison of MS/MS spectra (*above*: sample from West Giant Buddha, BMM201, *below*: cow glue, standard) (Kawahara, K., Fukakusa, S. and Nakazawa, T.: 12th Expert Working Group Meeting for the Safeguarding of the Cultural Landscape and Archaeological Remains of the Bamiyan Valley World Heritage Property, Orivieto, Italy (10–11 December 2013))

**Fig. 13.4** Photomicrograph of the polished cross-section of light blue is from sample no. BMM001 taken from the wall paintings of Cave N (a). Image taken in reflected normal light. A yellow resinous glaze was applied to the surface of drying-oil based paint layers which consisted of lead white and well levigated lapis lazuli blue. A translucent organic layer is seen between the earth render and the ground layer

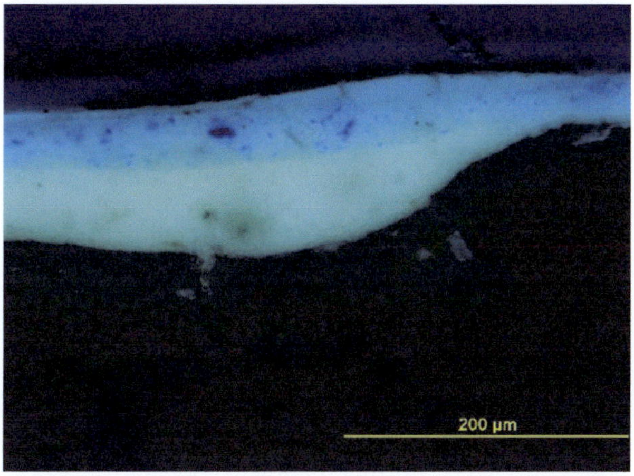


**Fig. 13.5** Photomicrograph of the same sample taken in reflected UV light

Archaeological objects carry a wealth of historical, technological and environmental information. In addition, they also carry aesthetic and monetary value. Explorers and archaeologists have unearthed countless archaeological objects and monuments across West Asia. Unfortunately, despite their inherent archaeological value, many of these objects and sites have been neglected owing to inadequacies in documentation, preventive and remedial care, maintenance, and storage environment, commonly stemming from a lack of resources among other things. Numerous precious archaeological objects continue to be found in formerly colonised countries, such as Egypt, Syria and Iraq. Whereas some select masterpieces were transferred long ago by colonial powers to museums in the West, many more remain in locally-based storage without adequate care. These items have, in many cases, lost their archaeological context and are in a poor state of conservation. In this sense, the immeasurable damage to these treasures also stems from a history of disequilibrium between West Asian countries and global powers.

Working with extraordinary objects originating from neglected sites in West Asia is like managing a second excavation—that is, one devoid of context or background information. Extensive scientific investigation is crucial for ensuring comprehensive conservation in such cases. Archaeological science helps to reveal technologies that rest beneath deteriorated materials, and may also function as a conservation tool by enabling diagnosis of the process and extent of deterioration so that appropriate measures can be taken for future protection. Many forms of material and technological information must still be studied since they are not yet fully understood or known, even in these modern times. In a way, archaeological science is able to shed light on demolished or destroyed cultural heritage once local inhabitants have sufficient latitude for seeking out cultural materials in the area. Such materials could be reassembled in future. However, without memories or recorded knowledge, they may not necessarily function as markers of an original cultural identity or common history.

## 13.5 Post-International Safeguarding Phase in Afghanistan

Personal involvement in UNESCO safeguarding projects in Iraq and Afghanistan has indicated that, following the end of the US-led war, ordinary approaches to cooperative work were not fully effective. Both people and government were too exhausted to safeguard cultural matters while immersed in post-war chaos and conflict. In the case of Iraq, the period of stability from the end of the war to the advent of ISIL was too short. Safeguarding damaged culture requires rehabilitation of households, education, livelihoods, respect and dignity for cultural identity, and probably tourism.

In the case of Bamiyan, it has been over 10 years since conservation activities began. Rehabilitation of the region led by international external initiatives did not provide sustainability. Although recent political and safety-related instabilities had posed problems for international cooperative work, locally-based autonomous activity oriented towards rehabilitation has been remarkable. Rapid development, such as that affecting infrastructure could not be made to follow a master plan for protecting cultural landscapes and sites. People must grapple with the more immediate and frequent dilemma of how to bring security into their environment and daily lives, at the same time. Therefore, a more flexible framework for rehabilitation would be preferable, to allow enough time to solve such problems (Taniguchi 2013).

Bamiyan is not a forgotten site. It remains living heritage in many ways—one that provides us with a continuous stream of new discoveries. The comprehensive approach taken with regard to the Bamiyan wall paintings and Buddha statues during the conservation project described above has given the world a rich variety of results providing testimony to manufacturing technologies of the time. This new knowledge has acted as an important key for raising local's interest in their heritage and reinforcing their own unity.

## 13.6 Cultural Identity and Tourism in Bamiyan

The town of Bamiyan has progressed rapidly in the development of its educational systems, social services, health care, and urban planning, among other things. It has transformed itself into a very different modern town compared to the village that it was in 2004. It is true that there is an increase in the ratio of illegal digging and land sharks (the price of the land around in Bamiyan has increased exponentially) due to the rapid development.

On the other hand, international aid organizations such as UNICEF have locally established infrastructures for education, medical care, and even transportation, such as the Bamiyan airport to allow for larger airplanes. It is particularly remarkable to see girls who used to engage in household work attend school. Students of Bamiyan University who specialise in tourism are keen on in-depth study of local history and cultural resource management (Fig. 13.6). Numerous cultural events,

**Fig. 13.6** Lectures in archaeology and conservation for students of Bamiyan University (2013)

including Bamiyan Culture Day, the Silk Road Festival, Tour de Bamiyan, and Bamiyan Ski Club, have been organised by the Bamiyan Tourism Association (BTA) and the Aga Khan Foundation with some aid from foreign NGOs (Figs. 13.7, 13.8, 13.9, 13.10, 13.11 and 13.12). Also, in 2013, a female student joined conservation work on wall paintings for the first time. This marks a dramatic change at Bamiyan University where students are encouraged to engage with their cultural heritage.

Interest in cultural heritage is also represented by the new Buddha statue erected at Bamiyan Airport. However, local enthusiasm for rehabilitation in combination with a reduction of pressure from the Taliban has prompted people to push quite far, quite fast: namely, in seeking to *rebuild* the destroyed Buddha statues.

On 7 June 2015, at the Cultural Day of Bamiyan, the millionaire Chinese couple Xinyu Zhang (张昕宇) and Hong Liang (梁红) put on a show illuminating the empty Buddha niche using 3D laser light projection technology—a feat worth approximately $120,000. The show was broadcast as part of a documentary programme called "On the Road" by the Chinese Internet television company YoukuTudou, Inc. According to the request letter by the Ministry of Information and Culture of Afghanistan, the government clearly wished to 'rebuild' one of the Buddhas using visualisation technology to draw international attention to Afghanistan's cultural heritage and preservation. The projection event received substantial attention from international media, and positive remarks from most locals (Figs. 13.13 and 13.14). There has been serious discussion over the last decade or more concerning whether the fragmented Buddhas should, in fact, be rebuilt. Although it is absolutely impossible to rebuild them using original fragments at their original locations, initial trials of rebuilding the foot-like foundation of the

**Fig. 13.7** The 6th Bamiyan Silk Road Festival in 2014 (Photograph: Abbas Khawari)

**Fig. 13.8** Tug-of-war competition at the 6th Bamiyan Silk Road Festival in 2014 (Band-e-Amir, First Afghanistan's National Park) (Photograph: Abbas Khawari)

East Buddha were executed at Bamiyan by ICOMOS Germany in response to locals' desire to rebuild.

It is true that the area in front of two Buddhas means a lot to locals since it is the only open public space available for them. It has always functioned as a space for people to assemble for occasions such as football games, festivals and political events. Therefore, (re)creating the cultural icons that have for so long stood as a common symbol could play an important and even essential role in re-creating the community; they may also enrich their community through tourism and other inter-

**Fig. 13.9** Female participants at the 7th Silk Road festival and SAARC ceremony in Bamiyan in 2015. According to the Governor of Bamiyan 50,000 people attended (Photograph: Mohammad Ali Shaida, Bamiyan Tourism Association)

**Fig. 13.10** The 7th Bamiyan Silk Road Festival in 2015 (Photograph: Latif Azimi, Bamiyan Tourism Association)

**Fig. 13.11** Over 100 participants for the cycling event, Tour de Bamiyan, in August 2014 (Photograph: Mohammad Ali ShaidaBamiyan Tourism Association)

**Fig. 13.12** 4th Afghan Ski Challenge event at Koh-e-Baba, Bamiyan, February 2014 (Photograph: Bamiyan Tourism Association)

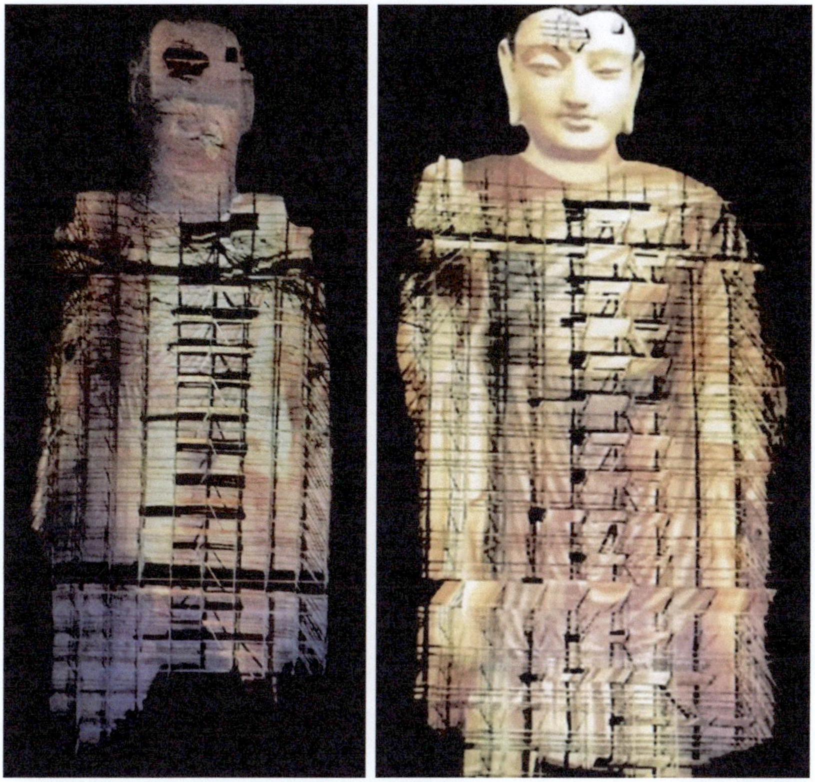

**Figs. 13.13 and 13.14** 3D laser imaging of Buddha before the destruction (*left*) and reconstructed image of the sixth century ACE Buddha (*right*) in the empty niche on 6 June 2015 (Photograph: H. Yasui)

national attention to a culture that the people of Bamiyan can be proud of (Figs. 13.15 and 13.16).

However, first and foremost, there ought to be carefully reached consensus on the significance of having Buddha figures within Afghan society. Reconstruction carries a high risk of criticism or possible attack by extremist actors. Moreover, while having statues may spur growth in tourism, thereby financially and culturally benefitting locals, it will also create disparity and possibly tensions between the local Hazara community and other ethnic groups outside the Bamiyan region. In reality, there has been increased violence against the Hazara since 2014 and many have been killed across Afghanistan and Pakistan (Fig. 13.17). Any decisions should be made with very careful forethought and sensitivity.

Today's engineering methods and tools may be able to provide technical solutions for reconstruction of destroyed statues. However, we face not only engineering challenges, but also ethical issues rooted in the conservation of cultural heritage. It

**Fig. 13.15** Miniature souvenir statues of the BamiyanBuddhas (Photograph: Latif Azimi, Bamiyan tourism Association)

**Fig. 13.16** Newly installed Buddha statue at the Bamiyan airport (Photograph: H. Yasui). June 2015

**Fig. 13.17** Street theatre performance of the slaughter of the Hazara presented in many provinces in Afghanistan including Wardak, Balkh, Zabul, Farah, Helmand, Herat and Sar-e-Pul (2015) (Photograph: Mohammad Ali Shaida, Bamiyan Tourism Association)

is essential to be aware of the tremendous ramifications any decision could have for local lives.

Cultural and academic value can be retrieved from demolished cultural heritage using scientific methodologies. The original physical form of destroyed cultural heritage could also be recovered either digitally or physically through conservation. However, historical collective memories and cultural unity and integrity cannot survive without the active participation and involvement of communities and people.

**Acknowledgements** I would like to thank the Ministry of Information and Culture of Afghanistan and the JCICC for providing me with the opportunity to carry out a conservation project inclusive of scientific analyses in Bamiyan, and to the Afghanistan National Museum's former General Director Omara Khan Massoudi, Abbas Khawari from Bamiyan Museum, Mohammad Ali Shaida and Latif Azimi from the Bamiyan Tourism Association, and Hiromi Yasui from Kyodo News for kindly sharing recent photographs from Bamiyan.

## References


Bahrani, Z. 2013. Babylon: A case study in the military occupation of an archaeological site. In *Of the past, for the future: Integrating archaeology and conservation, proceedings of the conservation theme of the 5th world archaeological congress, Washington, DC. June 2003*, ed. N. Agnew and J. Bridgland, 240–250. Los Angeles: The Getty Conservation Institute.

Brent, M. 1994a. Le grand pillage du Mali. *Le vif-L'Express* 2218: 48–52.

Brent, M. 1994b. The rape of Mali. *Archaeology* 47: 26–35.

Cather, S. 2010. Choices and judgment: The professional conservator at the interface. In *Conservation of ancient sites on the silk road, proceedings of the second international conference on the conservation of grotto sites, Mogao Grottoes, Dunhuang, People's Republic of China, June 28–July 3, 2004*, ed. N. Agnew, 22–32. Los Angeles: Getty Conservation Institute.

Crippa, C., R. Granata, and C. Margottini. 2013. Afghanistan, Buddhas of Bamiyan: Emergency consolidation of the cliffs and niches. In *Geotechnical engineering for the preservation of monuments and historic sites*, ed. E. Bilotta, A. Flora, and S. Lirer, 279–288. London: Taylor and Francis Group.

Gibson, M. 2008. The acquisition of antiquities in Iraq, 19th Century to 2003, legal and illegal. In *The destruction of cultural heritage in Iraq*, Heritage matters' series, ed. P. Stone and J.F. Bajjaly, 31–40. Woodbridge: The Boydell Press.

Haramoto, T. 2009. The destruction of cultural properties during ethnic conflicts-the cases of Bosnia and Herzegovina-. *International Public Policy Studies* 13: 127–140.

Harmanşah, Ö. 2015. ISIS, heritage, and the spectacles of destruction in the global media. *Near Eastern Archaeology* 78: 170–177.

Iwai, S. 2007. Radiocarbon dating and art-historical studies in Central Asian mural paintings, In *Mural paintings of the silk road: Cultural exchanges between east and west*, ed. K. Yamauchi, Y. Taniguchi, and T.Uno (Proceedings of the 29th annual international symposium on the conservation and restoration of cultural property, national research institute for cultural properties, Tokyo, January 2006, 54–59). London: Archetype Publications.

JCICC (ed.). 2004. *Protecting the world heritage site of Bamiyan: Key issues for the establishment of a comprehensive management plan 2004*. Tokyo: National Research Institute for Cultural Properties.

JCICC (ed.). 2007a. *Preliminary report on the safeguarding of the conservation of the Bamiyan site: 2005 – 4th and 5th Missions*, Recent cultural heritage issues in Afghanistan. Preliminary report series 1. Tokyo/Nara: Ministry of Information and Culture, Islamic Republic of Afghanistan/Japan Center for International Cooperation in Conservation/National Research Institute for Cultural Properties.

JCICC (ed.). 2007b. *Preliminary report on the safeguarding of the conservation of the Bamiyan site: 2006 – 6th and 7th missions.* Recent cultural heritage issues in Afghanistan. Preliminary report series 2. Tokyo/Nara: Ministry of Information and Culture, Islamic Republic of Afghanistan/Japan Center for International Cooperation in Conservation/National Research Institute for Cultural Properties.

JCICC (ed.). 2012. *Preliminary report on the safeguarding of the Bamiyan site 2009, 2010 – 9th and 10th missions*, Recent cultural heritage issues in Afghanistan preliminary report series 6. Tokyo/Nara: Ministry of Information and Culture, Islamic Republic of Afghanistan, Japan Center for International Cooperation in Conservation, National Research Institute for Cultural Properties.

Latour, B. 2002. What is iconoclash? Or is there a world beyond the image wars? In *Iconoclash: Image wars in science, religion and art*, ed. B. Latour and P. Weibel, 14–37. Cambridge, MA: The MIT Press.

Margottini, C. 2009. Recovery of the Buddha's niches and cliff in Bamiyan (Central Afghanistan) after the Taliban destruction of 2001. In *Landslides: Disaster risk reduction*, ed. K. Sassa and P. Canuti, 191–210. Berlin/Heidelberg: Springer.

Miyaji, A. 2002. *Bāmiyān, harukanari: ushinawaretaBukkyō bijutsu no sekai*. Tokyo: NHK Books.

Miyaji, A. 2006. The art-historical study on Bāmiyān and the radiocarbon dating. In *Radiocarbon dating of the Bamiyan mural painting*, Recent cultural heritage issues in Afghanistan, vol. 2, ed. K. Yamauchi, 131–144. Tokyo: NRICP and Nagoya University Museum, Akashi Shoten.

Petzet, M. (ed.). 2009. *The giant Buddhas of Bamiyan: Safeguarding the remains*, Monuments and sites series, vol. XIX. Berlin: ICOMOS-Hendrik Bäßler.

Rothfield, L. 2009. *The rape of Mesopotamia-behind the looting of the Iraq museum*. Chicago/London: The University of Chicago Press.

Shaheen, K. 2015, August 25. *Islamic state releases images said to show destruction of Palmyra Temple*. The Guardian. http://www.theguardian.com/world/2015/aug/25/islamic-state-images-destruction-palmyra-temple-baal-shamin-isis. Accessed on 14 Sept 2015.

Taniguchi, Y. 2007. Issues of conservation for the Bamiyan Buddhist mural paintings. In *Mural paintings of the silk road: cultural exchanges between east and west*, ed. K. Yamauchi, Y. Taniguchi, and T. Uno (Proceedings of the 29th annual international symposium on the conservation and restoration of cultural property, national research institute for cultural properties, Tokyo, January 2006): 144–151. London: Archetype Publications.

Taniguchi, Y. 2010. *Painting technology of Bamiyan Buddhist wall paintings*. Doctoral thesis, University of Tsukuba.

Taniguchi, Y. 2012a. Constituent material analysis of the Bamiyan Buddhist wall paintings in Central Asia (1): Using synchrotron-based μFTIR, μXRF/μXRD analyses. *Bulletin of National Museum of Japanese History* 177: 29–79.

Taniguchi, Y. 2012b. Constituent material analysis of the Bamiyan Buddhist wall paintings in Central Asia (2): Organic analysis using GC/MS and ELISA. *Bulletin of National Museum of Japanese History* 177: 80–105.

Taniguchi, Y. 2013. Conserving the Buddhist wall paintings of Bamiyan in Afghanistan: Practical issues and dilemmas. In *Art of merit: Studies in Buddhist art and its conservation*, ed. D. Park, K. Wangmo, and S. Cather. London: Archetype Publications.

Taniguchi, Y., and S. Aoki. 2005. Conservation proposal. In *Protecting the world heritage site of Bamiyan: Key issues for the establishment of a comprehensive management plan 2004*, ed. JCICC, 76–90. Tokyo: National Research Institute for Cultural Properties.

Yamauchi, K. (ed.) 2006. *Radiocarbon dating of the Bamiyan Mural paintings* (Recent cultural heritage issues in Afghanistan), Vol.2. Japan Center for International Cooperation in Conservation, National Research Institute for Cultural Properties, Japan, Nagoya University Museum. Tokyo: Akashi Shoten.

Yamauchi, K. (ed.) 2013. *Conservation of the Mural paintings of the Bamiyan Buddhist caves*, Documentation of the Bamiyan Site 3, Recent cultural heritage issues in Afghanistan, Supplement 6, National Research Institute for Cultural Properties, Tokyo.

# Chapter 14
# Conservation and Archaeological Science for the Protection and Understanding of Ancient Craftsmanship in Cultural Heritage

**Stavroula Golfomitsou, Thilo Rehren, and Christian Eckmann**

Archaeological objects are made of different materials, most of which are susceptible to deterioration. These objects represent the past in more than one way. They are material remains that can provide information about past civilisations and cultures, but they also represent the people who made or commissioned them, the people who used them, as well as the environment they were exposed to over hundreds or even thousands of years following their discard or deposition. All this information would be lost if the objects once excavated are not conserved, stored and studied properly. This chapter discusses how archaeological and conservation science can be used to study objects and reveal crucial information about the skills of the ancient craftsmen, information that is often hidden either deliberately by the maker of the object, or because of changes which happen after deposition, such as the formation of corrosion crusts or loss of material. As an example, the conservation and analysis of two statues from Egypt (of Pepi I), which were excavated more than a hundred years before their study and conservation was undertaken, are discussed. Finally, the chapter discusses how science can be a tool in better understanding the past and both the tangible and intangible heritage value these objects carry.

S. Golfomitsou (✉) • Th. Rehren
UCL Qatar, Hamad bin Khalifa University, Georgetown Building,
25256, Doha, Qatar
e-mail: s.golfomitsou@ucl.ac.uk; th.rehren@ucl.ac.uk

C. Eckmann
Römisch-Germanisches Zentralmuseum, Ernst-Ludwig-Platz 2,
55116, Mainz, Germany
e-mail: eckmann@rgzm.de



A. Tsuneki et al. (eds.), *Ancient West Asian Civilization*,
DOI 10.1007/978-981-10-0554-1_14

## 14.1 Introduction

The most important tangible evidence of the achievements of past civilizations lies in the artefacts that survive from these ancient cultures. These artefacts were skilfully made from a wide variety of materials, including natural materials such as shells, fibres, stone, wood and bone as well as man-made materials, such as ceramic, metal, glass and pigments. In archaeological excavations, the more robust materials survive, such as ceramic, metal, glass and stone as organic materials have often long since perished. Even when ancient objects have survived burial they are very susceptible to rapid deterioration once excavated, as seen for instance in the fading of the colours of the Terracotta Army figures shortly after their excavation, or the wide-spread 'bronze disease' which affects countless bronze objects in museum storerooms worldwide.

Even seemingly resilient materials such as metals deteriorate during their long exposure to harsh environmental conditions. Buried objects can reach equilibrium with their environment, especially if they are buried in relatively stable conditions; however, over hundreds or thousands of years in the soil most metals will corrode slowly but surely. Different materials survive better in different conditions, for instance metals are better preserved if buried in dry environmental conditions, while wood fares much better in a waterlogged environment, or when dry. It is important to realise that once excavated, the environment in which the objects are held changes dramatically, as they are now exposed to new conditions, often with daily and annual cycles of temperature, humidity and light exposure. As a result, the deterioration rate of almost all materials accelerates massively after excavation; in the case of metals this causes further changes in the mass and composition of corrosion products (Berducou 1996, 249).

As professionals in archaeological and conservation science, it is our responsibility to deal with this threat to the integrity of archaeological objects, which is a direct result of human action – excavation (Rehren 2002). First, we need to minimise the effects this change might have on the different materials objects are made of. Our aim is to physically preserve these valuable and irreplaceable objects as best we can. However, we also need to study and document them for future generations before they decay completely, and also for the benefit of people far away who cannot visit the original objects. This study, however, does much more than just document the physical object – it reveals the skill of the ancient craftsman, thus unlocking the intangible heritage that underpins all subsequent technical developments up to the modern day.

Deterioration and corrosion result in chemical and morphological changes, such as the loss of the original surface of an artefact through erosion, or hide it through the formation of encrustations covering the object. In advanced stages this can lead to the total destruction of an object. Conservation and conservation science aim at halting deterioration processes through either preventive or remedial methods. Preventive methods aim at modifying the environment an object is exposed to in order to halt or at least delay deterioration processes. Stable and low-energy environmental conditions, designed to match the requirements for each specific material, are central to this. Remedial methods on the other hand are applied directly to the object with the same aim, but also trying to reverse some corrosive effects. The development over recent decades of conservation as a profession and conservation science as a scientific field have helped us understand deterioration and corrosion processes better and develop treatments that ensure the long-term survival of archaeological objects. In archaeological objects, corrosion products are part of the biography of artefacts and need to be preserved to a certain degree while slowing down the pace of further corrosion as much as possible. Archaeological conservation aims to stabilise an object and where necessary restore its aesthetics, but does not aim to restore its functionality. This puts conservation in stark contrast to modern industrial practice which aims primarily at preserving the function and economic value of an object and sees the complete removal of any corrosion products as a necessary prerequisite before preventive measures are taken to stop the development of new corrosion.

In this chapter we argue that conservation has a twofold role to play in the management of cultural heritage. It not only manages the physical change materials undergo over time; it is also central to preserving and revealing the cultural information carried by the object. Slowing down deterioration is vital to preserve the object for future generations and for further research so we can study and understand the manufacturing skills of the ancient people that made these artefacts. These skills are often intentionally hidden beneath the surface of the object, hidden from the viewer so as not to distract from the aesthetic appeal of the finished artefact. The technical skill of the craftsperson in creating an object is as important and as remarkable as the aesthetic skill of the artisan finishing the appearance of the object to match the style and fashion of the age. The technical skills of craftspeople are an important part of the intangible heritage of every culture, but often overlooked in favour of the more artistic side of cultural heritage. However, the technical skills of craftsmen and women are at the root not only of the creation of every individual artefact, but of all scientific and engineering progress over the last 10,000 years. It is our privilege and duty as conservation and archaeological scientists to discover and cherish the achievements of our ancestors. The development of these skills over time is fundamental to our modern society, and those surviving archaeological objects are precious and rare manifestations of these skills. Many of these early skills are also immensely impressive, especially in those cases where people lacked the technical or engineering means and scientific understanding which we have today, but nevertheless excelled in their crafts which allowed them to construct objects and monuments of outstanding value, admired even today, such as the Great Buddha in Todai-ji in Nara, Japan.

One very early example of such skills is the case study presented below. Here, we demonstrate the level of information retrievable during conservation and through careful scientific study.

## 14.2 The Intangible Value of Everyday Objects

Objects have different meanings and values for different groups of people. Meaning and value change depending on the context the object is found in or displayed (Pye 2001). Objects represent different things and cannot be examined out of context.

However, the context, like its use, changes during the lifetime of an object. Objects are found in a specific context that represents a moment in their lifetime; however, this is not the whole story. For example, an object found in a tomb could have been used by the deceased during their life, or it could have been of symbolic or emotional value for the person buried, or it could represent a gift for the after-life. The object could have been imported from another site or country and might have been used by different people for other purposes before it was placed in the tomb. Evidence of these changes in use and function of an object are often preserved on its surface, and it is important that they are understood and be preserved for future generations. Conservation can reveal this information but certain steps, such as the removal of surface deposits or corrosion layers during cleaning, if done inappropriately could result in a loss of information or specific attributes and meanings the object has, thus affecting its authenticity (Pye 2001).

Everyday objects were not made to last forever but were typically manufactured for specific purposes, often using inferior materials. In addition, everyday objects do not usually survive well as they were heavily used and often carelessly discarded rather than carefully buried when they could not serve their purpose anymore. Also, in the past these objects were rarely collected during excavation. Often museum collections are mostly made of exceptional objects or artefacts made with good quality materials that managed to withstand the ravages of time. By their very nature, such exceptional objects are not representative of the past, but are exceptions. In contrast, there is a lot to learn about the past from everyday objects, which represent both the person who made them as well as the person they were made for. The vast majority of people in the past had access only to everyday objects; understanding the past lives of this 'silent majority' of our ancestors depends largely on the interpretation of these everyday finds. For us today, the technological skills, trade and other craft-related issues preserved in ancient objects are often easier to understand compared to the intangible social and religious values these objects carried in their time. However, studying the past through material culture is complicated and we should allow new technologies in the future to add to our interpretation of these findings. For this reason it is imperative that we preserve all the evidence an object carries, whether we can or cannot comprehend it.

## 14.3 Conservation and Archaeological Science

Conservation is not a new profession; evidence of mending of broken ceramics or worn metal artefacts is found even in antiquity. However, over the last century there have been significant changes in the professional practice and identity of this field, with conservators now regularly making use of scientific approaches in their investigation and treatment of an object (Muňoz Viňas 2005, 3–7). The involvement of scientists in the analysis of archaeological objects traces back to the late eighteenth century with the analysis of coins, glass and pigments. Already these first studies offered an insight into ancient technology (Winter 2005) even if they were driven primarily by curiosity and not necessarily by specific research questions. The development of instrumental analysis since the beginning of the twentieth century allowed more frequent use of new methods for the study of archaeological artefacts. The development of non-destructive, non-invasive methods has also helped in the use of analysis on a more regular basis in museums and heritage organisations.

One of the first conservation laboratories to open was the Chemical Laboratory of the Royal Museums in Berlin, with Dr Friederich Rathgen, a chemist, as its director. The laboratory was established in 1888 and focused on both the characterisation of archaeological objects and the development of conservation treatments (Clavir 2002, 10). The predominant definition of conservation scientists refers to professionals trained in natural sciences and applying their knowledge to the conservation of cultural heritage (Corbeil 2000). In addition to their original training, scientists need to have a clear understanding of the ethical principles of conservation in order to become conservation scientists (Price 2000).

Non-destructive, non-invasive methods refer to the methods where no sample needs to be removed from an object for its analysis. Techniques such as portable X-Ray Fluorescence (pXRF) are commonly used these days in museums. XRF offers elemental chemical analysis of inorganic materials, most importantly metals, ceramics and glass. However, such non-destructive surface analysis might not be the best way to provide the necessary information. Selective sampling from an object can give important information related to the technology of the object, its original surface, alterations during use and burial, and more. Taking samples from archaeological objects comes with a number of restrictions, as sampling should only be allowed when necessary and with a clear question in mind. Samples should be taken from areas which are not easily visible (such as for example a broken edge, a base, etc.), but which also might not be representative of the object. All these restrictions should be discussed and taken into consideration when discussing the potential of scientific analysis.

In many cases, however, removal of a sample for study and analysis can be justified and highly informative, even necessary, in the interest of the object and our understanding, documentation and preservation of cultural heritage. This is similar to the study of a small tissue sample taken by a medical professional, which is often essential for a proper diagnosis and the development of an effective treatment plan. Even if a patient has already died, an autopsy is often necessary to learn about the disease and prevent further deaths.

## 14.4 Constructive Analysis

An example on the information value of small samples is the study of a series of entirely corroded and badly preserved iron artefacts from the Early Iron Age site of Lefkandi, east of Athens in Greece. The corrosion of iron is often totally destructive, particularly in coastal environments rich in salt and with variable humidity. A restoration of such corroded artefacts is often impossible, and even large prestige items such as swords or daggers from important burials can be completely destroyed even before they were excavated. However, in many cases the microscopic study of small

**Fig. 14.1** Fragment of a corroded sword blade found attached to other objects, Lefkandi, Greece

mounted samples enables us to see the final stages of the life histories of these objects, and even to reconstruct their original quality.

Figure 14.1 shows a fragment of a totally corroded sword blade from a tomb from Lefkandi, corroded together with other fragments of iron weapons. This cemetery dates to the tenth century BC, the very beginning of the Iron Age in this part of the world, and the iron objects are among the earliest known of their kind. This early date raises the question whether people used the soft bloomery iron, or whether they had already mastered the use of the iron-carbon alloy, steel. Normally, metallographic investigation of a small piece of metal would answer this question. However, as is evident from the cross section there is no metal left (Fig. 14.2), even though the original outline of the blade is still visible. Detailed study using optical and electron microscopy then showed that the original metal was mild steel with about half of the volume having been pearlite (Figs. 14.3 and 14.4). This is remarkable since the finds are among the oldest known iron artefacts, and the use of steel at such an early time was not expected. In addition, several of the finds from this cemetery had a thick crust or layer of magnetite outlining the original surface (Figs. 14.5 and 14.6). This is a very specific structure and formation normally known as hammer scale (Dungworth and Wilkes 2009, their Fig. 3), and different from the magnetite which can form during corrosion. Normally, hammer scale is found as tiny flakes scattered on the floor of a smithing workshop, and characterised by a very thin layer of haematite (darker grey in Fig. 14.6) on a body of a massive magnetite sheet. Here, however, the hammer scale is still attached to the artefact, which showed that the blade had been part of a cremation pyre where the scale formed, but was not then hammered – at which point the brittle scale would normally flake off and be removed from the object.

**Fig. 14.2** Cross section of the iron blade. Evidence of the limits of the original surface are still preserved within corrosion layers

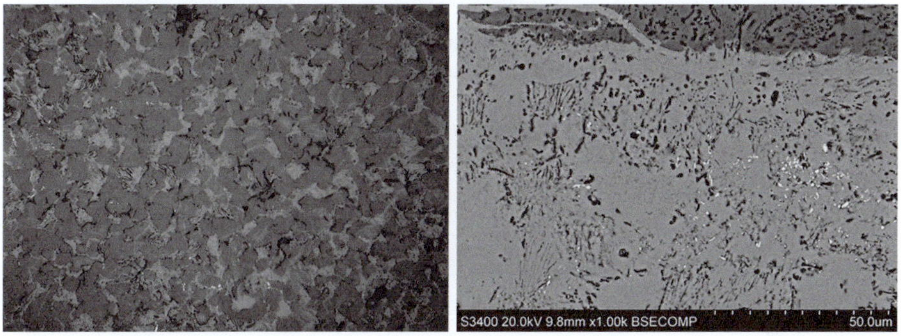


**Figs. 14.3 and 14.4** SEM images of the sword showing the structure of mild steel with pearlite phases

The point here is that although, in this particular case, there is no metal left and the object is in such a fragmented and corroded state that no conservation or restoration will make it fit for exhibition, there is still valuable information that can be retrieved from it – through invasive analysis. With no well-preserved sword from such an early period it is difficult to take full cross sections for analysis, let alone different sections from different parts of the blade. Here, however, it is possible to do this and to reveal the elaborate shape to which the sword had been worked (Fig. 14.2), as well as identify that steel had been used in its manufacture (Figs. 14.3 and 14.4), and that the cutting edge of the blade was not different in composition from the main body (as it often was in later swords). Identifying the cremation history of the sword also helps to interpret what metallographic structure is present, regardless of whether the object is well preserved or totally corroded. The fact that the object was included in the pyre would have resulted in a full annealing of the iron, removing most traces of any work hardening or other thermal treatment that

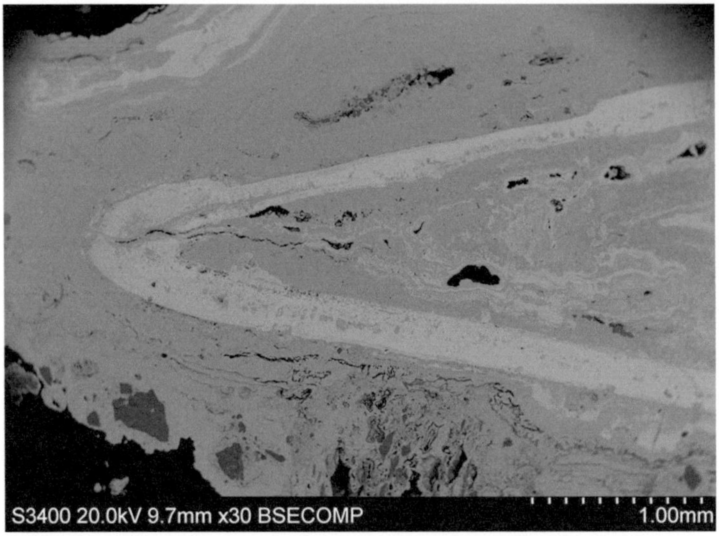


**Fig. 14.5** SEM image of the mild steel sword from Lefkandi. A thick layer of magnetite signifies the limits of the original surface of the blade (magnification 30×)

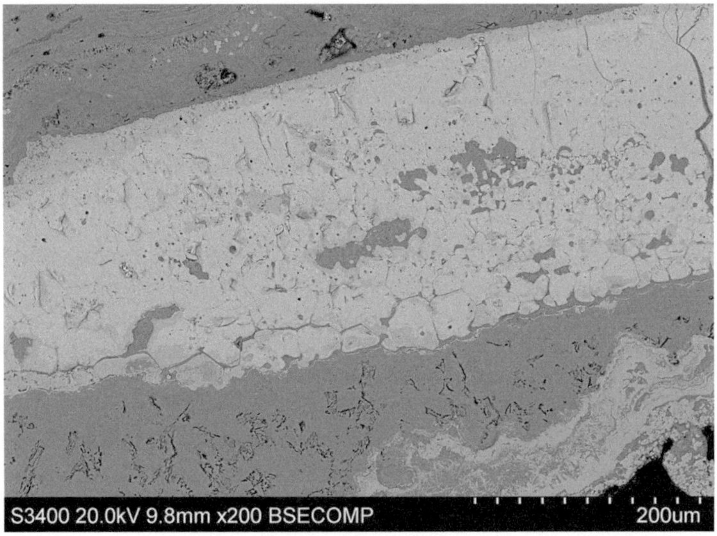


**Fig. 14.6** SEM image of magnetite signifies the limits of the original surface of the blade (magnification 200×)

the object might have had during its 'functional' life. Thus, it would appear now much softer and less professionally made than it might have been in reality, possibly leading to the wrong interpretation that the sword was a purely decorative item made specifically for the burial, when in fact it might have been a fully functioning weapon taken from the dead warrior's possessions. This is why we prefer to refer to this kind of study as constructive analysis, in contrast to the more often used 'destructive analysis' whenever the removal of a sample is involved. We maintain that from a cultural heritage viewpoint, this study does not destroy anything which isn't already totally destroyed, but instead creates a tremendous amount of knowledge which otherwise would be unobtainable.

Another excellent example of this is the analysis of the two Pepi I statues discussed below, again creating and constructing knowledge in relation to the technology, life and deterioration of an object.

## 14.5 The Statues of Pepi I

For more than one hundred years visitors to the Egyptian Museum in Cairo have looked in fascination at two statues that were excavated in 1897 by the archaeologists Quibell and Green at Hierakonpolis in the south of Egypt. Both statues were part of the numerous finds of a 2-year archaeological campaign during the course of which remains of a temple district of a predynastic settlement were excavated. Shortly after the archaeologists began their excavation, they made initial sensational discoveries. Quibell reports: …"*the sand was not clean, and on digging further we found at a depth of 1m two legs of a life-size copper statue. Below and beyond the thighs the face appeared, and the left forearm stood up on one side… When the two legs of the statue had been removed, the torso below was seen to be incomplete, the hips being missing… The metal statue was badly corroded and in a poor condition*". On the chest and adhering to it by corrosion, lay a crumpled sheet of copper with an embossed inscription on it. As can be seen from the excavators' records the thin and fragile copper sheet broke into scores of pieces during an attempt to separate it from the torso. The plate was originally part of a wooden base, which had been separated before the statue was buried in the soil. However, despite the poor condition of the metal sheet its inscription allowed the excavators to identify the statue as the VI. Dynasty King Pepi I. (2268–2228 BC), making this by far the earliest life-size metal statue known. After the statue was retrieved and during an attempt to clean out the hollow trunk, the torso was found to contain another smaller statue in a similar state of disassembly. The torso was brought to the Cairo Museum with the small statue still inside and there both figures were carefully separated, assembled and put on exhibition. Figure 14.7 shows the statue after initial restoration in 1900 (left) and Fig. 14.8 shows the statue again in 2003 after the recent conservation treatment (right). In view of the size of the statues, the material they were made of and the manufacturing techniques applied, they can be considered as unique examples of Pharaonic works of art in a very early period of human history.

Due to their corroded state it was not possible to see how they were produced, and their manufacture has been the subject of discussion ever since their discovery. At the time it was assumed that they were the earliest cast copper figures known to mankind. Unfortunately, a continuing process of decay has been visible since they

**Fig. 14.7** King Pepi I statue, before conservation

were first exhibited. Most worryingly, there were structural changes which became evident even on the surface. Within the last 100 years air pollutants and inconsistent humidity and temperature levels affected the statues and corrosion started anew, resulting in the chemical transformation of the remaining metal into a green powder and surface flaking.

In 1996 a cooperative project was initiated aiming at the preservation and the technological examination of these two unique pieces of art. A major study of the manufacture was done by Christian Eckmann and Ms S. Shafik from the Egyptian Museum, Cairo. The initial assessment showed that the statues suffered from extensive corrosion as well as a great number of cracks caused by the pressure of the statute's own weight and a continuously decreasing stability due to corroding material. The original surfaces were covered with bulky corrosion crusts and burial accretions that neither allowed scholars to appreciate the way these statues were made, nor to see the original surface of the object and its decoration (Fig. 14.9). However, the original surface is of great importance, because it bears valuable information. The features of interest include surface decoration, inscriptions, plating, inlays or organic coatings. Therefore, the major task of any cleaning process is to expose this layer without destroying it – bearing in mind that cleaning is an irreversible process.

**Fig. 14.8** King Pepi I statue, after the 2003 conservation treatment

**Fig. 14.9** Head of King Pepi I statue before conservation

**Fig. 14.10** Head of King Pepi I statue after the 2003 conservation treatment

The cleaning process of both statues was carried out exclusively by mechanical means, such as the careful use of scalpels, dental and ultrasonic tools. With the help of a binocular microscope, the original surface was revealed (Fig. 14.10). The cleaning process of the small statue took more than a year, while the large life-size statue required about 2 years. The careful cleaning process revealed traces of gilded gesso layers which had been hidden from view showing the elaborate decoration applied to the figures. Figure 14.11 illustrates the hand before treatment and Fig. 14.12 shows the hand after cleaning. The artisan's attention to achieve a naturalistic modelling is remarkable and particularly evident is the care given to the modelling of the hands and feet. Almond-shaped eyes made from coloured stones, plastically worked out eyebrows tending back into flaring cosmetic lines and accentuated lips give the statues an almost life-like air. These findings together with the examination of technological details significantly increased the archaeological information obtained from these objects.

Painstaking inspection and re-assemblage of thousands of fragments (Fig. 14.13) enabled the reconstruction of the pedestal on which the large figure originally stood, including the inscription which identified it as King Pepi. X-ray imaging of the statues revealed the way they were made, and fundamentally changed our understanding of Old Kingdom metallurgical skills (Fig. 14.14). It became clear that even the large parts were made of individual metal sheets riveted together, and

**Fig. 14.11** Detail of the hand of King Pepi I statue before conservation

that no casting was involved in their production (Fig. 14.15). Although the X-rays showed that the metal had been mineralised to a large extent through corrosion processes, samples were taken from loose fragments to understand how these statues were manufactured. At low magnification the laminar structure typical of hammered metal became visible in a cross section. The samples displayed a surface-parallel microstructure and further, the flattened ends showed that chisels had been used for cutting the copper sheets. Based on the metallographic evidence and the material composition we can now say that the statues were definitely not cast.

Of particular interest is the way the different sheets were joined together and whether there was a core (e.g., wood) inside the hollow figure, or not. The metal sheets are joined with rows of copper nails placed remarkably close together, forming long seams along the arms and legs. Microscopic examination in combination with SEM imaging illustrates in detail how the rivets were placed (Fig. 14.16).

The use of nails for joining copper sheets implies the existence of some core material into which the nails had been hammered. For this reason, one might easily be inclined to suppose that the ancient artisans used a fully articulated wooden core as a support over which they hammered the copper sheets. Although the theory of a carved wooden core was widely accepted among Egyptologists, there was no evidence to prove it. With the exception of two tiny charcoal remains adhering to the

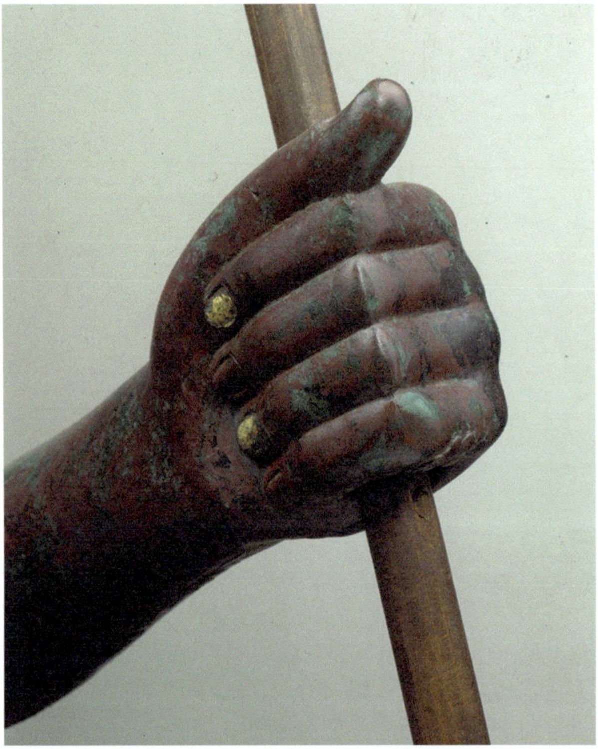

**Fig. 14.12** Detail of the hand of King Pepi I after the 2003 conservation showing gilding decoration on the nails

inside of the small statue, no traces of organic material were detected, not even in the spaces between the nails which would be the most likely. Reconsidering the circumstances of the discovery, the use of a wooden core seems rather unlikely. As already mentioned above, the archaeologists found the statues dismantled with the smaller statue placed inside the torso of the larger one. If there had been a wooden core, how could the parts have been stored inside? In fact, a carved wooden core installed inside the statue would have presented considerable technical difficulties as hammering copper sheets into shape needs access from both sides and strains the crystalline structure of the metal, eventually rendering it too hard for further work. At this point it must be annealed in the fire – a process which allows the crystals to reshape in a new unstressed arrangement, and thus restores the malleability of the metal, but which cannot be done with a wooden core inside.

Further observations strengthen the scepticism with regard to a carved wooden core. As mentioned above, the hip and the crown of the large statue were missing and the fact that the base of the torso and the tops of the thighs show unbroken edges allows us to conclude that the missing parts of the body must have been made of some material other than copper – most probably wood. In contrast to the characteristic feature of joining two copper sheets with rows of nails placed closely together, those areas of the statues where the metal met the missing parts show perforations at much wider distances. In view of manufacturing this appears most comprehensible, as one would join the metal to the wooden elements using only a few nails.

**Fig. 14.13** Conservator Ms S.Shafik working on the fragmented pedestal

**Fig. 14.14** X-ray imaging of the leg of the small Pepi I statue showing the way the metal sheets were riveted together

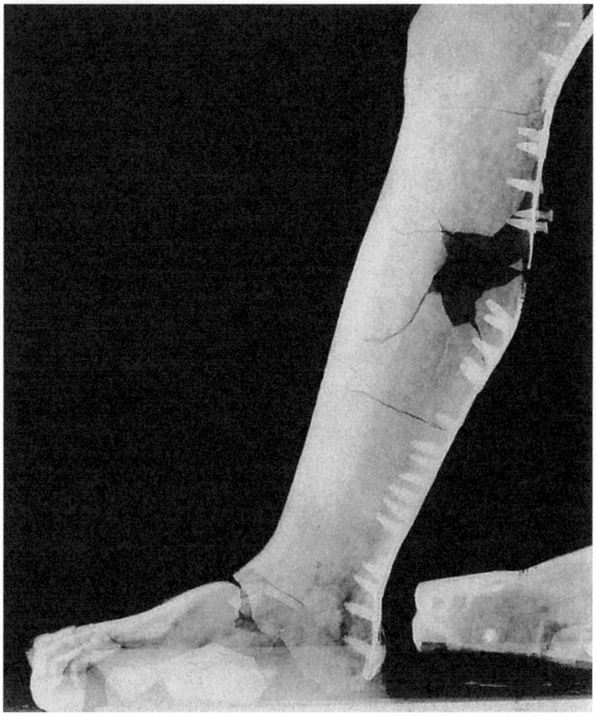

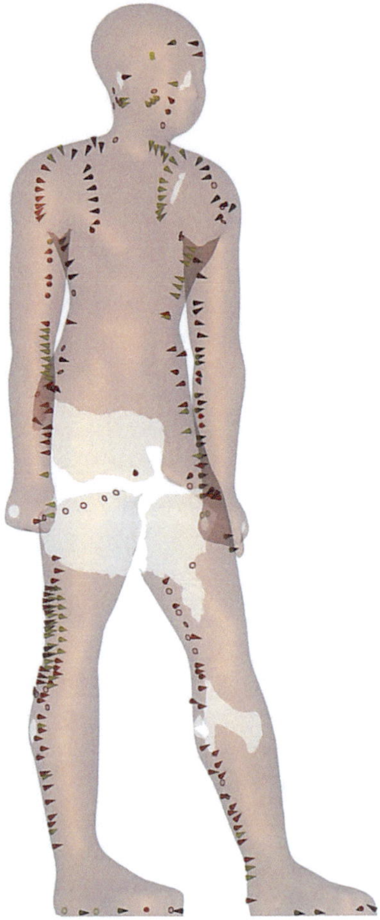

**Fig. 14.15** Illustration of the manufacturing technique used for the Pepi I statues

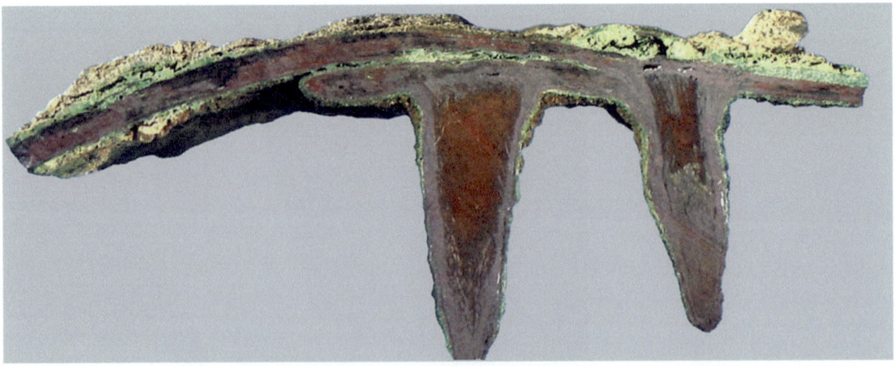

**Fig. 14.16** Detail of sample taken from the Pepi I statue demonstrating the way the metal sheets were riveted together

Even though a wooden core is unlikely it seems that the Pepi statues must have had some kind of internal armature. There are good reasons to suppose that this armature together with the wooden parts of the statue such as the missing kilt, which also served as a link between the thighs and the torso, had been removed before the figures were dismantled, then packed one into the other and finally buried in the temple.

## 14.6 Conclusion

Several issues can be highlighted from these case studies regarding the role of conservation science in cultural heritage. Documentation during excavation is extremely important, as the second case illustrates. The statues were excavated a 100 years before their actual conservation, and some missing pieces and additional information are yet to be found. However, through scientific examination of the statues during conservation we were able to reveal a plethora of information hidden underneath corrosion layers. Outwardly, there were clear remains of the earlier decoration of the figurines with gilding and inlayed stones that provided a glimpse of the original splendour of the statues. But even underneath this decorated and carefully smoothened surface we were able to discover unexpected information about the skills of the early metal-smiths more than four millennia ago. Not only did they know how to cast and hammer thin sheets of copper metal, but they had the skill to bend and shape it in a very naturalistic manner to create a life-size statue, doubtlessly including numerous cycles of hammering and annealing. The example shows the importance of preparation and planning in an excavation, to include proper documentation of the conditions an object is found in, and any initial conservation undertaken on site. Deterioration is accelerated after excavation even if an object is stored in a museum, and we need to minimise the effect of the shock the materials suffer during and after their discovery, due to the sudden change in environmental conditions. There are numerous examples of objects which either got damaged or were totally destroyed during excavation; a famous example includes the coloured surfaces of the Terracotta Army warriors, which faded within a few hours after excavation. Conservation science provides the tools we need to protect and document these objects. Preparations should start prior to the excavation and continue during excavation until an object is properly stored in a stable and protective environment. Although the excitement of the discovery often prohibits professionals from following the right steps, we should always act having the protection of our cultural heritage in mind, even if this restricts immediate access to the object.

Scientific investigations are key to understanding not only the technological developments but also the people and their skills over the years. Different techniques are available including: non-invasive, non-destructive and invasive, which involve sampling of the original material. Each technique has different things to offer and one should always start from the least invasive and move to more invasive techniques only if required. The consideration here is similar to that taken by a medical doctor – the patient needs treatment, and while no harm must be done to the patient it is sometimesnecessary to remove a small piece of tissue to determine the nature of the problem. There are clear principles we need to follow regarding sampling to ensure it will not be destructive to the original object; but if done properly and professionally it will generate a lot of information, not only about the physical state of the object but also about the skills of the people who made it, thousands of years ago.

Thus, through analysis the intangible values of craftsmanship and skills can be revealed. Conservation methods help us prevent further deterioration of the material, and scientific analysis and other investigations together give us an insight into the life and knowledge of the people in a specific period. Material science is vital to completing the picture of the past and understanding how people lived and worked. Although modern engineering allows mass production of everyday objects, the individual objects from the past carry the craft of the people within them.

In this paper we aimed to show how materials science and conservation science work in tandem to better understand ancient civilisations and their achievements; how to present ancient objects without misguiding the observer; and how corrosion covers but also preserves evidence, and is part of the biography of an object. Thus, only the combination of art history, excavation and post-excavation documentation and technical and scientific analysis allows a full reconstruction of ancient skills. Most of these investigations are best done during the conservation treatment of an object, giving conservators a unique opportunity and responsibility to reveal the full picture of our heritage, including the amazing intangible skills that manifest themselves in ancient artefacts and everyday objects.

**Acknowledgements** We are very grateful for the generosity of our colleague S. Shafik in allowing us to present this work on this occasion, and for the organisers of the conference for their kind invitation for us to participate. We are also grateful to Dr Thomas Loughlin for editing the paper.

## References

Berducou, M. 1996. Introduction to archaeological conservation. In *Historical and philosophical issues in the conservation of cultural heritage*, ed. N.S. Price, M.K. Talley, and A. Melucco Vaccaro, 248–260. Los Angeles: The Getty Conservation Institute.

Clavir, M. 2002. *Preserving what is valued: Museums, conservation, and first Nations*. Vancouver: UBC Press.

Corbeil, M.C. 2000. Training options for conservation scientists in university postgraduate curricula for conservation scientists. In University postgraduate curricula for conservation scientists: Proceedings of the international seminar, Bologna, 26–27 November 1999. Rome: ICCROM..

Dungworth, D., and R. Wilkes. 2009. Understanding hammerscale: The use of high-speed film and electron microscopy. *Historical Metallurgy* 43: 33–46.

Muňoz Viňas, S. 2005. *Contemporary theory of conservation*. Oxford: Elsevier Butterworth-Heinemann.

Price, C.A. 2000. Are conservation scientists real scientists? In *University postgraduate curricula for conservation scientists: Proceedings of the international seminar*, Bologna, 26–27 November 1999, 75–80. Rome: ICCROM.

Pye, E. 2001. *Caring for the past: Issues in conservation for archaeology and museums*. London: James & James.

Winter, J. 2005. The State of the Field: Overview. In *S cientific examination of art: Modern techniques in conservation and analysis*, 3–11. Proceedings of colloquium at the National Academy of Sciences, 19–21 March 2003. Washington, DC: The National Academies Press.

## Further Reading

Artioli, G. 2010. *Scientific methods and cultural heritage*. Oxford: Oxford University Press.

Brothwell, D.R., and A.M. Pollard. 2001. *Handbook of archaeological sciences*. Chichester: John Wiley & Sons.

de Waal, E. 2010. *The Hare with Amber eyes: A hidden inheritance*. London: Vintage.

Eckmann, C., and S. Shafik. 2005. *Leben demHorus Pepi: Restaurierung und Technologische Untersuchung der Metallskulpturen des Pharao Pepi I.* aus *Hierakonpolis. Mainz: Verl. des Römisch-Germanischen Zentralmuseums.*

Martinón-Torres, M., and Th. Rehren. 2008. *Archaeology, history and science: Integrating approaches to ancient materials*. Walnut Creek: Left Coast Press.

Miller, D. 2010. *Stuff*. Cambridge: Polity Press.

Rehren, Th., and E.B. Pusch. 2005. Late bronze age glass production at Qantir-Piramesses, Egypt. *Science* 308: 1756–1758.

Shortland, A.J., I.C. Freestone, Th. Rehren, and M.S. Tite. 2009. *From mine to microscope: Advances in the study of ancient technology*. Oxford: Oxbow Books.

Vitobello, M.L., and Th. Rehren. 2009. *A Quest for authenticity: Authentication methodologies for metal artefacts based on material composition and manufacturing techniques*. London: Institute for Archaeo-Metallurgical Studies IAMS, UCL Institute of Archaeology.

# Part V
# Synthesis

# Chapter 15
# Some Observations and Future Directions

**Timothy P. Harrison**

> *Archaeology is perhaps the best tool we have for looking ahead, because it provides a deep reading of the direction and momentum of our course through time: what we are, where we have come from, and therefore where we are most likely to be going.*
>
> *(Ronald Wright, Massey Lectures 2004: 56)*

## 15.1 Introduction

At a time of heightened conflict and turmoil in the Middle East, a deeper understanding of the region's civilizational foundations is both timely and critically important. As noted by Akira Tsuneki in his introduction, the Middle East today is seen to be an intractable source of conflict and political instability, resistant in particular to Western attempts to institute Western ideals of liberal democracy and political economy. Typically characterized as a clash of civilizations, as most famously quipped by Samuel Huntington (1993, 1996), between an Islamic Middle East and a (largely) Judeo-Christian West, their engagement is all too commonly seen to be at an impasse, and ultimately irreconcilable.

Yet, as the theme of this volume appropriately highlights, the Middle East, or ancient West Asia, has been the source of many of the great civilizational accomplishments the modern world now enjoys. It witnessed the emergence of the first sedentary human communities, the domestication of plant and animal life, the innovation of multiple craft technologies, urbanization, the first formal social, political and religious institutions, complex writing systems and bureaucracies, specialized large-scale economies, and the first interregional commercial and political networks. Put simply, ancient West Asia was the birthplace of civilization. Moreover, the roots of Western civilization are inextricably linked to this ancient West Asian civilizational history. The study of this civilizational history is therefore critical to an understanding of the cultural and institutional foundations of the West, and ultimately of the modern world.

T.P. Harrison (✉)
Near and Middle Eastern Civilizations, University of Toronto,
4 Bancroft Avenue, 2nd floor, Toronto M5S 1C1, Ontario, Canada
e-mail: tim.harrison@utoronto.ca



A. Tsuneki et al. (eds.), *Ancient West Asian Civilization*,
DOI 10.1007/978-981-10-0554-1_15

I believe such a deepened understanding will not only provide much needed insight into the root dynamics that continue to inform the social, economic and political forces that drive the modern era, it will also create an opportunity to foster a comparative intellectual framework within which more collaborative intercultural interaction and civilizational dialogue might be possible. In what follows, I will first summarize briefly the principal topics explored in this volume, highlighting a number of recurring themes, followed by a review of several prevailing methodological perspectives. I will then identify areas of contemporary concern where I believe the deep-time perspective of ancient West Asian civilizational history can make a direct contribution, and propose how we might implement such an approach.

## 15.2 Themes in Ancient West Asian Civilization

The papers in this volume cover a wide geographical and chronological range, but they nevertheless focus on a number of key topical themes. The first concerns the environment, and more specifically the constraining role of geography and the geology of the Middle East. Hisada examines the geodynamics of West Asia, and demonstrates the richness of the natural sources in the region, which have enabled ancient populations to develop their innovative cultures. Aghanabati raises the question of the need for a geology of culture, and emphasizes the importance of developing a relationship between earth science and indigenous cultures, especially within the context of the study of the cultures of the ancient Middle East. Similarly, Anma and Maruoka note the fundamentally interdependent relationship between climate change and anthropogenic forces, and demonstrate the profound role this has played in shaping both the physical and cultural landscapes of the region. The paper by Yagi and Ohsumi, meanwhile, illustrates the formative impact seismic activity can have in shaping, and constraining, the civilizational accomplishments of a community.

The second section shifts the time-scale from geological eras to human developmental epochs, with a focus on the origins of agriculture, one of the great transformative ‘events’ in human history. Tanno and Maeda review the archaeobotanical evidence, which points to the broad based nature of early plant domestication, the result of widespread experimentation over an extended period of time, and argue for the need to move beyond early theories that have emphasized a sudden, chronologically and geographically delimited ‘origin’ for early agricultural development. Similarly, regarding the domestication of animals, Uerpmann and Uerpmann emphasize that this transformation was not the result of a single, sudden event, or discovery, but was the product of extended experimentation, as communities recognized the economic potential of animals, and shifted their strategies and the nature of the relationship from predator to protector. They also note the important distinction between domestication and the subsequent commodification of animals that occurred as human communities interacted more intimately with them over time, and discovered an ever expanding range of ways that animals might be exploited.

Pyrotechnology represents the third major human innovation that was introduced during this transformative era. Yalçın stresses the close relationship between the introduction of pyrotechnology and the 'domestication' of metal and the development of metallurgy, noting the evidence of metal use in the Neolithic as early as the ninth millennium BCE. He further emphasizes the impact this innovative development had on other technologies, in particular craft industries involving materials such as clay, stone, and organic remains. These early technologies had a profound effect on early farming communities, and ultimately began to introduce a wide range of tools that these communities were able to exploit in their efforts to produce greater resources for their members.

From the emergence of the first human sedentary communities in the Epipaleolithic and Neolithic periods, we move to urbanization and the birth of the first cities, the third great transformative episode in human history. As documented by Ur, the first truly urban communities emerged during the fourth and third millennia, and did so not only in southern Mesopotamia, but as archaeological excavations are increasingly discovering, also in Syria and northern Iraq, or northern Mesopotamia. The emergence of these urban communities also coincided with the rise of social complexity, the development of formal social, religious and economic institutions, and ultimately of state formation. What defines a city, and when, or what, precisely is the threshold at which villages and towns structurally become cities, is crucial to our understanding of these transformative social, economic, and political developments, and ultimately of the root causes that triggered this transformation.

The invention of writing, as described by Taylor, Ikeda and Yamada, was one of the great technological achievements of this era, and helped to secure the urban transformation that came with the creation of the first cities and the rise of social complexity. Writing, in the form of the cuneiform script, created virtually unlimited possibilities for communities to manage and document the increasingly complex social, economic and political relationships they now faced. While the first written records indicate these early writing systems were concerned primarily with tracking the movement of economic goods, and therefore functioned essentially in an administrative capacity as accounting systems, a broader range of genre quickly followed, including the production of literary and religious texts, reflecting the expanding awareness of the technology's ability to communicate complex concepts and ideas in both space and time. How widespread literacy became remains a hotly debated topic, though it likely remained limited to a small scribal class until the introduction of the alphabet almost two millennia later.

Religion, and more generally, the role of ideology in shaping formal, institutionalized belief systems, was another critical outgrowth of social complexity, and the increasingly hierarchical configuration of the institutions that came to structure ancient Mesopotamian society. Yet, as Gabbay observes, Mesopotamian religion, and the religious traditions of West Asia more broadly, were also remarkably diverse and multi-dimensional in character, incorporating a wide range of regional pantheons, theologies, and ritual practices. Indeed, their fundamentally syncretistic character is perhaps their most defining feature. Writing and iconography played critically important roles in conveying the conceptual sophistication of the myths and belief systems these communities created as they explored and navigated the liminal world between the human and divine realms. The potential to bring together archaeological, historical and textual perspectives on complex human social institutions like religion represents one of the more exciting developments in the study of ancient West Asia today.

The final set of papers review important developments in conservation science, and the complicated issues that have emerged in the face of the recent onslaught of cultural heritage destruction in the Middle East. Taniguchi discusses the role and limitations of conservation science, in light of the harsh realities of the ongoing destruction of cultural heritage and the looting of archaeological sites. She stresses the difficult technical challenges involved, but also the complicated ethical issues rooted in the conservation of cultural heritage. Golfomitsou, Rehren and Eckmann, meanwhile, survey the remarkable advancements in archeological science and conservation in recent years, and emphasize the importance of coordinating the research efforts of archaeology (in particular archaeometry) with conservation science. Using metal statues of the Egyptian Old Kingdom pharaoh Pepi I and his son as a case study, they illustrate how we can advance our knowledge of the cultural traditions of a community, while also gaining insight into the technologies it used to produce artistic expressions of its cultural heritage.

## 15.3 The Importance of a Deep-Time Perspective

In 2014, at a conference on the theme of this volume, Professor Yutaka Tsujinaka, Director of the Institute for Comparative Research in the Human and Social Sciences at the University of Tsukuba, articulated a vision for a collaborative research environment that might facilitate the study of the world's civilizations in comparative perspective, and in doing so, contribute to the development of trans-national methods in humanities and social science research, while also advancing the study of global civilization and human history. As home to the first great civilizational experiment, West Asia must form a core program area in this enterprise. A key goal of this effort will be to identify effective strategies for overcoming the conflict and violence that has historically characterized interaction between civilizations, and a deep knowledge of ancient West Asian civilization will be integral to any mutual understanding such an undertaking might ultimately achieve.

This commitment to a deep cultural and historical understanding, and the belief in its relevance to issues of contemporary concern, particularly within the conflict zones of the Middle East, is a refreshing perspective that contrasts sharply with the essentialist, or 'orientalist', approaches that have dominated European and North American academic traditions. It also provides a unifying conceptual focus to the papers presented in this volume, and to the West Asian Civilization Studies disciplinary program Akira Tsuneki has proposed in his introduction. The value of multiple, especially non-Western, perspectives will be critically important to the ultimate success of this effort.

## 15.4 Methodological Perspectives

While the value of a deep-time perspective might be clear, and its relevance to issues of contemporary concern now increasingly acknowledged implementing an approach that incorporates this perspective remains a challenge. More specifically, how might a deep understanding of civilizational history and human social development contribute to the production of strategies aimed at addressing pressing contemporary issues? A wide range of theoretical perspectives and scholarly traditions have attempted to address this question. While this is not the place for a thorough review, much of the literature can be grouped according to a number of common themes.

The first of these might be called “long-term lock-in” theories, as Ian Morris has aptly labelled them (2010, 13). Their unifying view is that a critical determinative factor, or set of factors, has shaped the developmental trajectories of every civilization, from the earliest civilizations of ancient West, East and South Asia, to more recent medieval and modern exemplars. I have already mentioned Huntington’s ‘clash of civilizations’ perspective (1993, 1996). Other similarly-minded theorists include the geographer Jared Diamond, whose widely read, *Guns, Germs, and Steel: The Fates of Human Societies* (1997), sought to isolate the key environmental and technological factors that have determined the fate of the world’s civilizations, and most recently, the historian Niall Ferguson. In his provocative, *Civilization: The West and the Rest* (2011), Ferguson identifies six “killer apps” (competition, science, the rule of law, modern medicine, consumerism, and the work ethic) that he believes gave Western civilization a competitive advantage over other, primarily Eastern, civilizations. These theorists assume the superiority of a particular tradition, and then attempt to identify the key factor(s), whether environmental or non-environmental, that helped it achieve the civilizational accomplishments for which it is known.

A second group might be characterized as “short-term accident” theorists, following Morris’ categorization (2010, 18–21). These theorists (e.g., Frank 1998; Pomeranz 2000; Goldstone 2009), often using Western civilization as an example, argue that civilizations follow their own unique developmental trajectories, with minimal impact from their specific environmental (spatial) and chronological (temporal) settings. Historically contingent, or ‘accidental’, forces, agents and/or events, in their assessment, are the critical element(s) that give shape to the distinctive character of each civilizational experience.

Most recently, Ian Morris (2010, 2013) has introduced a quantitative approach that seeks to measure the social developmental performance of a civilization, essentially a society’s ability to get things done in the world, as he puts it, permitting—at least in principle—the possibility of comparing civilizations across space and time. Morris defines social development as, “the bundle of technological, subsistence, organizational, and cultural accomplishments through which people feed, clothe, house, and reproduce themselves, explain the world around them, resolve disputes within their communities, extend their power at the expense of other communities, and defend themselves against others' attempts to extend power" (2013, 5). His social development index attempts to measure and quantify four key traits, or performance indicators (energy capture per capita, social organization, war-making capacity, and information technology), during the life cycle of the great civilizations that have come and gone over the past 15,000 years of human history. In so doing, this approach also creates the possibility for a more systematic comparative analysis of civilizational history. Morris emphasizes that this social development index is intended to provide a *measure* of history, and therefore should not be viewed as explanatory, nor does it seek to advance a value-laden agenda.

Whatever we might think of these different theoretical approaches, one perspective they all share is the strong belief that the study of civilizational history has contemporary relevance, and that what we learn about the past has direct bearing on our ability to understand and anticipate the potential directions in which our modern world—in all of its civilizational complexity—is heading, both in the present, and in the future.

## 15.5 Implementation and Future Directions

The value of studying civilizational history notwithstanding, there remains the challenge of implementation. I would advocate for a targeted approach that focuses on key thematic subject areas where a deep-time perspective, particularly within the West Asian context, might be able to contribute directly to issues of contemporary concern.

The most obvious subject area is the environment, and more specifically environmental and climatic change, both subjects of broad and increasing concern. Several of the papers in this volume have effectively demonstrated the central importance of the environment in shaping the social and cultural worlds in which we live. They have also highlighted ways that a study of the environment might contribute directly to current debates about anthropogenic causes of environmental degradation, and ultimately even of climate change. In particular, the ability of archaeology to contribute finely stratified cultural sequences spanning thousands of years, especially time ranges that coincide with the rise of social complexity and urbanization, and then correlate these sequences with tightly calibrated paleoenvironmental proxy data drawn from cores and other sources of climate data, introduces an unparalleled time depth over which to examine the impact of human activity within a clearly defined spatial and temporal context. In addition, as chronometric sequences are refined, for example, through the wiggle-matching of radiocarbon evidence with dendrochronology, it will soon be possible to measure this time depth in increments bordering on the scale of annual events, possibly even historically documented events.

The time depth and granular resolution that a carefully excavated cultural sequence can provide also affords the possibility of examining the long-term consequences of social and economic strategies deployed by human communities over time. A deep-time study of a community, for example, can offer insight into that community's subsistence strategies, how these strategies evolved over time, and their long-term impact on the nutrition and health of the community, its local environment and resource base. Collaborations between archaeologists, paleo-ecologists and biochemists, meanwhile, have recently begun to extend this focus to the microbial communities found within archaeological sites, where the long-term view afforded by multiple period settlements offers the opportunity to examine the evolving biochemical structure of these communities, what role human agency might have played in advancing this micro-evolutionary process, and even the prospect of discovering previously undocumented biochemical compounds and microbiological life forms.

A third subject area is ethnic and sectarian conflict, certainly a pressing concern within the West Asian or Middle Eastern context. As archaeologists, we believe many of the contemporary issues that drive the region's conflicts today are deeply rooted in dynamics and patterns that extend into the distant past. The role of memory, cultural identity, and place are inextricably embedded within the social fabric of a community, and ultimately help to foster a sense of meaning and order. A knowledge and understanding of this cultural history, and the underlying dynamics that have given it shape, offers the possibility of bridging cultural and ethnic divides and building understanding between communities, including a respect for difference, the value of diversity, and the existential importance of collaboration.

On a broader inter-regional or transnational scale, a deepened understanding of the civilizational history of the Middle East, or ancient West Asia, has the capacity to help mitigate the 'clash' of civilizations that threatens to engulf the region today. As Christina Luke and Morag Kersel argue in their prescient book, *U.S. Cultural Diplomacy and Archaeology: Soft Power, Hard Heritage* (2013), on a very pragmatic level, this should include an acknowledgement of the important cultural diplomatic role archaeologists can play in the course of conducting long-term field research within communities. Investing in local relationships and community engagement should be encouraged as an integral part of field research, and recognized as an opportunity to build understanding across often deep cultural and civilizational divides.

How might we advance the contribution of a deep-time perspective in these subject areas? Space does not permit a detailed response to this question, but an implementation strategy should involve at least three key elements: (1) clearly articulated research questions and/or agenda, (2) a collaborative research approach, and (3) a program of fully integrated data analysis. The importance of the first two elements is generally acknowledged, and increasingly even reflected in actual practice, the third considerably less so.

Over 150 years of active archaeological fieldwork and data collection have generated a staggering range and volume of information about the cultures and civilizations of ancient West Asia. As we have seen, this research has documented the first sedentary communities, the domestication of plant and animal species, the first formal social, political and religious institutions, writing systems and bureaucracies, specialized large-scale economies, and the first interregional commercial and political networks. Thus, as we have argued, the region offers an unparalleled laboratory for studying the long-term growth and development of human communities and their interaction with the natural environment.

Yet, it must also be said that we have failed thus far to achieve a truly collaborative and integrated analytical framework, or environment, that incorporates the full evidentiary range produced by this uniquely rich cultural legacy, limiting its contribution to a deeper understanding of the root causes and effects of socio-cultural change. While it might be accurate to claim that archaeology has emerged as one of the most dynamic and innovative disciplines in the humanities and social sciences, employing a truly interdisciplinary, collaborative approach and a continually expanding array of analytical research tools, together with an overwhelming volume of digital data, it is also true that attempts to store and analyze this growing body of data have foundered on issues of access, data compatibility, integration and analytical capability. Ultimately, creating a collaborative environment capable of facilitating such analyses will be essential if we are to achieve the full research potential of our discipline, and contribute substantively to the important intellectual concerns raised in this volume.

## References

Diamond, J. 1997. *Guns, germs, and steel: The fates of human societies*. New York: Norton Press.
Ferguson, N. 2011. *Civilization: The west and the rest*. New York: Penguin.
Frank, A.G. 1998. *ReOrient: Global economy in the Asian age*. Berkeley: University of California Press.
Goldstone, J. 2009. *Why Europe? The rise of the west in world history, 1500–1850*. Boston: McGraw-Hill.
Huntington, S. 1993. The clash of civilizations? *Foreign Affairs* 72(3): 22–49.
Huntington, S. 1996. *The clash of civilizations and the remaking of world order*. New York: Simon and Schuster.
Luke, C., and M.M. Kersel. 2013. *U.S. cultural diplomacy and archaeology: Soft power, hard heritage*. New York: Routledge.
Morris, I. 2010. *Why the West Rules—For Now: The patterns of history, and what they reveal about the future*. New York: Farrar, Straus and Giroux.
Morris, I. 2013. *The measure of civilization: How social development decides the fate of nations*. Princeton: Princeton University Press.
Pomeranz, K. 2000. *The great divergence: China, Europe, and the making of the modern world economy*. Princeton: Princeton University Press.
Wright, R. 2004. *A short history of progress*. Toronto: Anansi Press.

# Conclusion

We hope that this book has demonstrated a wider view of the interrelations between the environment and innovative human cultures created in ancient West Asia, as well as the significant influence of the West Asian cultures on the formation of modern civil societies.

The geological environment in West Asia is very distinct, being characterized by the presence of an orogenic belt, comprising the Zagros Mountains and the Taurus Mountains. These mountains and the adjacent areas surrounding the plains of the so called fertile crescent are prolific in raw materials, including hard lithic material suitable for the production of stone tools, metals such as copper, tin and silver, as well as oil that has been exploited intensively only in the modern age. The rich resource areas facilitated the development of the innovative material cultures that developed throughout the Stone, Bronze and Iron ages.

In addition to the super-rich resource conditions, the location of West Asia played an important role. If it is taken into consideration that our ancestors came from East Africa 120,000–60,000 years ago and went via the roundabout route of the Sinai or the Arabian Peninsula, they could have reached this orogenic belt and built up their Paleolithic cultures *in situ* using the abundant natural resources. Furthermore, they would have found desirable living conditions represented by many limestone caves and abundant food. They would also have discovered the useful radiolarian limestone. Thus, it appears that the natural resources in West Asia provided them with a good basis for the formation of an innovative ancient culture.

More than 10,000 years ago, a number of cultural innovations, including the domestication of plants and animals, metallurgy, urbanization and writing, were achieved earlier in West Asia than in other areas of the globe. People encountered the wild ancestors of modern wheat, barley, beans, sheep, goat, pig and cattle in the territory of West Asia. They also discovered the orogenic belt produced useful resources for human beings, such as copper and iron. West Asia is a land that has been blessed by God. However, domestication of plants and animals, innovative technologies and social innovations were not an invention attained in a short period but a process taking an extended period of time. One of the reasons for its long



A. Tsuneki et al. (eds.), *Ancient West Asian Civilization*,
DOI 10.1007/978-981-10-0554-1

duration is that all of these innovations appeared autonomously in West Asia without any outside influence. Each innovation has a long preceding developmental history and we can trace each of these stages. This allows us to grasp the innovations that took place in West Asia from the viewpoint of a deep-time perspective. The observation of what occurred in West Asia in antiquity may lead us to understand the long process of transition in human culture and societies.

The writing system, which was invented along with the oldest example of urbanization in southern Mesopotamia around 3000 BC, diffused widely into various surrounding regions. Thus, an exceptionally spaced and well-developed literate and civilized world was created in ancient West Asia, as early as the first half of the second millennium BC. It was characterized by a complicated political-administrative system, elaborate laws, dynamic economic activities, a rapid and sophisticated communication system, as well as the development of various metaphysical ideas. All of those cultural innovations caused an increase in social complexity, and social complexity in its turn accelerated cultural innovation, and *vice versa*. As a result of this chain-reaction in many aspects of society, West Asia established itself as the origin of civilization, which has continued in many ways as the basis of the way of life for many people on the globe today.

It seems that the outstanding dominance of the civilized lifestyle of the modern world reflects its efficiency as a survival strategy of population groups. The effective organization of power granted by civilization enabled people to face various difficulties stemming from natural or human causes, and to be successful in competition with their neighbors. This competition has had positive and negative results. It has sometimes yielded mutual prosperity, or otherwise caused bitter struggle and wars between different population groups, often seriously damaging some of them. Civilization in itself is probably not harmful for our world, if it is well controlled and properly directed. There is useful information and lessons from extensive data about what ancient West Asia experienced during its extremely long period of antiquity, which could help mitigate the destructive antagonism unfortunately prevailing throughout the globe in the modern age. It is a worthwhile endeavor to create a new trend in the *Study of Ancient West Asian CIvilization*, particularly focusing on the history of West Asia prior to Islam. This is a promising research field and provides us with opportunity for reflection. It is useful not only for eliminating the harmful prejudges against the Middle East, but also to supply a useful basis for pondering the future of human civilization, in the light of current problems.

Editors